T0253646

Cambridge Elements ≡

Organizational Response to Climate Change
edited by
Aseem Prakash
University of Washington
Jennifer Hadden
University of Maryland
David Konisky
Indiana University
Matthew Potoski
UC Santa Barbara

FIGHTING CLIMATE CHANGE THROUGH SHAMING

Sharon Yadin
*Yezreel Valley College School of Public
Administration and Public Policy and University
of Haifa Faculty of Law*

CAMBRIDGE
UNIVERSITY PRESS

Shaftesbury Road, Cambridge CB2 8EA, United Kingdom

One Liberty Plaza, 20th Floor, New York, NY 10006, USA

477 Williamstown Road, Port Melbourne, VIC 3207, Australia

314–321, 3rd Floor, Plot 3, Splendor Forum, Jasola District Centre, New Delhi – 110025, India

103 Penang Road, #05–06/07, Visioncrest Commercial, Singapore 238467

Cambridge University Press is part of Cambridge University Press & Assessment, a department of the University of Cambridge.

We share the University's mission to contribute to society through the pursuit of education, learning and research at the highest international levels of excellence.

www.cambridge.org
Information on this title: www.cambridge.org/9781009256261

DOI: 10.1017/9781009256230

First published 2023

A catalogue record for this publication is available from the British Library.

ISBN 978-1-009-25626-1 Paperback
ISSN 2753-9342 (online)
ISSN 2753-9334 (print)

Additional resources for this publication at www.cambridge.org/yadin_resources

Cambridge University Press & Assessment has no responsibility for the persistence or accuracy of URLs for external or third-party internet websites referred to in this publication and does not guarantee that any content on such websites is, or will remain, accurate or appropriate.

Fighting Climate Change through Shaming

Organizational Response to Climate Change

DOI: 10.1017/9781009256230
First published online: July 2023

Sharon Yadin
Yezreel Valley College School of Public Administration and Public Policy and University of Haifa Faculty of Law

Author for correspondence: Sharon Yadin, sharon@yadin.com

Abstract: This Element contends that regulators can and should shame companies into climate-responsible behavior by publicizing information on corporate contribution to climate change. Drawing on theories of regulatory shaming and environmental disclosure, the Element introduces a "regulatory climate shaming" framework, which utilizes corporate reputational sensitivities and the willingness of stakeholders to hold firms accountable for their actions in the climate crisis context. The Element explores the developing landscape of climate-shaming practices employed by governmental regulators in various jurisdictions via rankings, ratings, labeling, company reporting, lists, online databases, and other forms of information sharing regarding corporate climate performance and compliance. Against the backdrop of insufficient climate law and regulation worldwide, the Element offers a rich normative and descriptive theory and viable policy directions for regulatory climate shaming, taking into account the promises and pitfalls of this nascent approach as well as insights gained from implementing regulatory shaming in other fields.

This Element also has a video abstract: www.cambridge.org/yadin_abstract

Keywords: climate change, regulation, shaming, corporate compliance, public law

ISBNs: 9781009256261 (PB), 9781009256230 (OC)
ISSNs: 2753-9342 (online), 2753-9334 (print)

Contents

Online appendices for this publication are available at
www.cambridge.org/yadin_resources

1 Introduction: Shame and Climate Change

In an era in which governments are grappling with climate change, could regulation by government-initiated shaming of corporations help meet the challenge? In a recent survey conducted by Yale University, most respondents said that they are willing to engage in consumer activism by punishing companies that contribute to climate change, but that they do not know which companies to punish (Leiserowitz et al., 2021b). Many respondents said that they would like to engage in such climate activism, but that no one has ever asked them to. This Element studies a nascent approach to climate-change regulation, titled "regulatory climate shaming" (RCS), which enables regulators to name and shame companies in order to exert public pressure on these companies to cut emissions and adopt climate-friendly policies.

Regulatory climate-shaming schemes have begun to emerge in various forms and jurisdictions worldwide. For example, the US Environmental Protection Agency (EPA) has recently launched a database that enables users to view data on companies' greenhouse gas emissions in maps, charts, and graphs and to compare emission trends over time.[1] The Swedish Energy Agency now requires companies to place labels on fuel pumps, displaying company-specific climate-impact ratings for different fuels (Swedish Energy Agency, 2021). The UK Environment Agency is naming all the companies that have breached climate laws and regulations in the past year on its website, with details of the infringements.[2] And the Israeli Ministry of Environmental Protection publicly scores and rates factories and companies in a league table, based on climate and environmental performance.[3]

Both public shaming and climate change feature prominently in today's public discourse, and the idea of a regulatory tool that uses one to address the other has recently emerged as a novel combination of these concepts. Consequently, scholarship on climate shaming is now beginning to develop in the behavioral and social sciences and in the humanities. However, the research literature on shaming largely discusses climate change only as a secondary issue to more general environmental concerns. Additionally, the discussion usually revolves around various types of shaming actors and targets, including individuals, NGOs, countries, and the media. Scholarship dealing with mandatory environmental disclosure also has limited relevance because this practice is mostly focused on providing information to support consumer decision-making, rather than on shaming companies into compliance by utilizing the social and economic

---

[1] https://ghgdata.epa.gov/ghgp/main.do.

[2] https://data.gov.uk/dataset/13c0893a-049a-4608-9f9b-7f268a71f15a/climate-change-civil-penalties.

[3] www.gov.il/en/Departments/publications/reports/environmental_impact_index_annual_reports.

power of various stakeholders. It also largely deals with particular environmental issues rather than climate change.

Thus, the intersection of climate-change regulation and government shaming of corporations remains largely underdeveloped. This Element aims to fill this gap by developing a theory of RCS, hoping to pave the path for meaningful climate regulation advances worldwide, and to open new research avenues in this field.

I use the term "regulatory climate shaming" to refer to information conveyed to the public by government regulators on harmful corporate behavior that is contributing to climate change, with the aim of inducing corporations to comply with climate laws, rules, and regulations, and also to adopt voluntary climate norms. This Element's primary mission is to examine whether RCS should and could become a viable tool in the fight against climate change. Thus, it offers both a descriptive and normative theory for RCS as well as policy recommendations for its use in practice.

The Element will explore such questions as: How do regulatory shaming (RS) theory, climate-change law, regulation and governance literature, and environmental disclosure scholarship support the conceptual framework of "regulatory climate shaming"? What role can shaming play in the current regulatory landscape to address the climate crisis? What can we learn from shaming strategies that are already being deployed in the environmental regulation arena (which I will refer to in this Element as "regulatory eco-shaming"), and from RS in the health sector, for the formulation of sound climate policies? What are the characteristics of existing RCS schemes in various jurisdictions in the United States and in Europe? Which RCS strategies might work best in the near future? What are the main concerns and opportunities presented by RCS, and how can policymakers mitigate these concerns and maximize such opportunities? And can shaming be justified as a legitimate regulatory tool in the fight against climate change?

As a basis for developing the concept of RCS, this section will first look at several building blocks that may be more familiar to the reader: climate-change regulation (Section 1.1), climate shaming (Section 1.2), and shaming more generally (Section 1.3). These are intended to provide a broad perspective for examining the use of climate shaming of companies as part of governmental regulation – which is the focus of this Element – by also discussing various other actors and aspects of law, regulation, and governance pertaining to the topic. Section 1.4 will then briefly outline the Element's intended contribution.

1.1 Climate-Change Regulation

Climate-change regulation is currently one of the world's greatest challenges. It involves efforts on local, national, and international scales to mitigate global warming and its current and predicted extreme impacts on weather patterns and human lives. Generally, climate change refers to systemic long-term changes in climatic elements, such as temperature, precipitation, and wind (Dessler, 2021).

It is now well established that since the industrial revolution, the earth's temperature has risen markedly, mostly due to the burning of fossil fuels such as oil, coal, and natural gas (Maslin, 2021). This process releases greenhouse gases – primarily carbon dioxide (CO_2) into the atmosphere, warming the globe through a "greenhouse effect" (Archer & Rahmstorf, 2009). As a result, the eight years from 2015 to 2022 have been the warmest on record, and the global mean temperature in 2022 was around 1.15 °C above pre-industrial levels (WMO, 2023). The World Meteorological Organization predicts a 50:50 chance of the increase in global mean temperature reaching the 1.5 °C threshold in the next five years (WMO, 2022). Without immediate large-scale reductions in greenhouse gas emissions, global warming is predicted to climb to 2 °C above pre-industrial levels by 2040 (IPCC, 2021; IPCC, 2022b).

These changes pose a severe threat to air and water quality, biodiversity, and natural ecosystems (Dessler, 2021; IPCC, 2022a). Yet climate change is far from being merely an environmental issue, and its implications go well beyond changes to weather. It also holds dramatic implications for public health, food and housing security, infrastructure integrity, economic stability, national security, and various other fundamental aspects of our lives (Dessler, 2021; Future Earth, 2022). Extreme heatwaves, fires, storms, droughts, and floods are predicted to lead to increased water shortages, hunger and malnutrition, spread of infectious diseases, migration, conflicts over resources, poverty, and mortality (Maslin, 2021; McDonald, 2021).

These phenomena are already being experienced around the globe and are predicted to escalate in the near and far future (IPCC, 2021). In fact, the number of extreme weather events has increased fivefold over the past fifty years, causing some two million deaths, economic losses totaling more than $3.5 billion (WMO, 2021b), and a worrying increase in the number of climate refugees (Wennersten & Robbins, 2017). Against this background, it is not surprising that the UN secretary-general has recently referred to the situation as a "code red for humanity" (UNFCCC, 2021) and a "highway to climate hell" (van der Zee & Horton, 2022).

Climate change and its impacts have been known to the scientific community since the nineteenth century, yet it was not until recent decades that they attracted public and political attention (Archer & Rahmstorf, 2009; Dessler & Parson, 2019). Since the 1980s, climate-change regulation has been introduced on an increasing scale at the international, national, and subnational levels (Dessler & Parson, 2019). In the remainder of this section, I review these levels of regulation from a broad-brush perspective in order to underscore the dire need for effective climate regulation.

At the international level, several landmarks can be pointed out, chief among them are the 1992 Rio agreements, the 1997 Kyoto Protocol, and the 2015 Paris Agreement (Dessler & Parson, 2019). Notably, one of the 1992 Rio agreements, the United Nations Framework Convention on Climate Change (UNFCCC), serves as the parent treaty to subsequent international climate agreements (Carlarne et al., 2016). These agreements, achieved by some 150–200 nations in various UN summits, have evolved over time, from adopting general principles and vague obligations to setting concrete targets, most importantly for the reduction of greenhouse gas emissions (Benoit, 2022).

The conventional standards of current international climate regulation include keeping global warming well below 2 °C above pre-industrial levels (preferably 1.5 °C), reaching significant reductions in greenhouse gas emissions by 2030, and achieving net-zero greenhouse gas emissions by 2050 (Burck et al., 2021: 23; Dessler & Parson, 2019: 32; IPCC, 2022b). These standards are mostly based on the scientific reports of the Intergovernmental Panel on Climate Change (IPCC), the UN body tasked with assessing the science related to climate change.

Other central topics in international climate regulation include financial and technological assistance to developing countries (which was at the center of the COP27 summit in Sharm el-Sheikh), state pledges on deforestation, and the adoption of renewable energy technologies (Meckling & Allan, 2020). For example, during the 2021 COP26 summit in Glasgow, more than 100 countries pledged to halt deforestation by 2030, and around fifty states committed to a transition away from coal-generated power in the 2030s and 2040s (COP26, 2021).

Climate-change regulation at the national level has developed both as a derivative of international climate regulation and independently of it (Huang, 2021; Scotford et al., 2017). European Union member states have also developed climate law and regulation in accordance with EU legislation. Indeed, in recent years many states have passed climate-change mitigation laws, which address the root causes of climate change (such as coal-generated power) and seek to reduce their scope and impact (Burck et al., 2021: 23; European Environment Agency, 2022; Huang, 2021; World Bank, 2020). States are also advancing policies of climate-change adaptation, focused on providing better responses to current and expected impacts and implications of climate change, such as natural disasters, mass migration, and financial instability (Mayer, 2021; McDonald & McCormack, 2021; UNEP, 2022). The Grantham Research Institute's Climate Change Laws of the World database contains more than

2,400 laws and policies from some 200 countries on topics such as carbon pricing, low-carbon energy, industry emissions, fossil-fuel restrictions, deforestation, low-carbon construction and transportation, and natural disaster risk management.[4]

Many of these climate laws, rules, regulations, orders, decisions, programs, and guidelines are initiated, devised, implemented, and enforced by national administrative regulators, such as regulatory agencies and governmental ministries. For example, environmental agencies set greenhouse gas emission standards for vehicles and aircraft, implement programs to promote renewable fuels, and propose regulation to reduce emissions in the fossil-fuel sector (European Environment Agency, 2022; Freeman, 2020; World Bank, 2020). Other national regulators – in fields such as energy, transportation, health, planning, agriculture, security, commerce, and finance – also take part in climate-change regulation.[5] While international climate regulation is usually directed at countries (though corporations are also starting to engage in international climate agreements), governmental climate regulation tends to target corporations, facilities, businesses, industries, markets, and sectors (McDonald & McCormack, 2021).

Generally, government climate regulation harnesses a range of different types of tools (European Environment Agency, 2022; Gupta et al., 2007). These include various limitations, standards, permits, and prohibitions; cap-and-trade systems, which limit companies' permitted emissions through allowances and enable companies to purchase and sell unused allowances; disclosure schemes, which require that information on emissions and climate action is reported and publicized; voluntary public–private programs, which usually aim to achieve standards that transcend compliance with legally binding obligations ("beyond-compliance") (Hsueh, 2020; Hsueh & Prakash, 2012; Potoski & Prakash, 2009); regulatory agreements with companies and industries, which may address compliance or commitments to go "beyond-compliance" and typically include some form of regulatory leniency or commitment; and a variety of subsidies, financial incentives, charges, and taxes, which are worth mentioning here even though they are sometimes considered nonregulatory instruments (Fankhauser et al., 2010).

Climate regulation tools can be categorized, among other ways, according to their level of coerciveness: for example, pollution output requirements that are imposed via regulatory permits, rules, and regulations are generally considered hard, mandatory, command-and-control-style regulation; while other tools, such as regulatory agreements and disclosure schemes, are generally considered

---

[4] climate-laws.org.

[5] See, for example, Columbia University's US Climate Regulation Database, https://climate.law.columbia.edu/content/us-climate-regulation-database.

forms of soft regulation (Hsueh & Prakash, 2012). However, both hard and soft climate regulation tools may be based on administrative, criminal, or civil sanctioning, such as civil penalties and fines. For example, regulatory agreements, which companies can choose to enter voluntarily, often include provisions for the imposition of penalties upon infringement (Hsueh, 2020). Similarly, failure to produce or publicly present a building climate rating (a type of disclosure scheme) can result in penalties.

Climate regulation is also being conducted on the subnational level, by local governments and municipalities. These bodies advance, for instance, greenhouse gas reduction policies and energy efficiency schemes (such as "green building" policies), using various types of regulatory tools – similar to those deployed at national levels – applied via local laws, codes, ordinances, and the like (Moffa, 2020; Sørensen & Torfing, 2022). These also incorporate varying degrees of coerciveness, using hard and soft regulatory approaches and legal styles, and they are often directed at local businesses.

Yet, by and large, climate regulation at all three levels of governance[6] has produced disappointing results (Dessler & Parson, 2019; IPCC 2022b; Lyster, 2016). Countries are lagging behind their Paris Agreement goals, and even if the 2021 Glasgow COP26 pledges are fulfilled, the earth's temperature is expected to rise well above the 1.5 °C threshold (CAT, 2021; IEA, 2021; UNEP, 2021). While the COVID-19 pandemic resulted in a slight decrease in greenhouse gas emissions in 2020 (Le Quéré et al., 2021), 2021 has seen a noticeable rebound (IEA, 2021), continuing an unmistakable trend of emission growth over recent decades (UNEP, 2021).

Some attribute the failures of international climate law to a lack of enforcement mechanisms (Huggins, 2021), while others underscore the lack of participation of major emitting countries in UN Conference of the Parties (COP) summits and international agreements, alongside the highly politicized and consensus-based nature of the process, which involves dozens of nations (Genovese, 2020). Other explanations focus on emission targets being overly optimistic, unattainable, and set for up to three decades in the future (Burck et al., 2021: 24), as well as on the language of commitments being too soft and vague (Lyster, 2016).

National climate regulation is also considered insufficient, as some countries are only now beginning to legislate climate laws while others are still lacking any real legally binding domestic frameworks for climate mitigation and adaptation (IPCC, 2022b: ch. 5; Scotford et al., 2017; UNEP, 2022). Some researchers point to national climate policies that are legislated but not implemented *de facto*, or

---

[6] Alongside international, national, and subnational climate regulatory schemes, the private sector has also developed climate self-regulation mechanisms. These will be discussed briefly in Section 2.2, though generally, this subject is beyond the scope of this Element.

which include merely aspirational statements that affect greenhouse gas emissions only marginally (Eskander & Fankhauser, 2020). Many consider these regulatory failures to be the result of the fossil-fuel industry's efforts to actively deny climate change and thwart regulatory endeavors (see Section 2.2). Still, there is clearly a regulatory momentum on the national level across jurisdictions.[7] The subnational level of climate regulation also shows great promise (Moffa, 2020), though as it tends to thrive under climate regulation deficiencies at the national level (Carlarne, 2019), it might dwindle as the current momentum of national regulation continues.

Against this background, there seems to be a consensus that innovative new policies are desperately needed on the climate-change front (Carlarne et al., 2016; Coen et al., 2020; Dessler & Parson, 2019; IPCC, 2022b). There also currently appears to be a considerable degree of openness to implementing innovative regulatory tools at the national and subnational levels of climate regulation, and increasing opportunities to do so (IPCC, 2022b; Leiserowitz et al., 2021c).

1.2 Climate Shaming

Generally, "climate shaming" refers to the act of publicly denouncing or condemning individuals, business organizations, and countries for acts, omissions, and decisions that contribute, on a large or small scale, directly or indirectly, to climate change. The concept is most closely associated with "flight shaming," "meat shaming," and other types of "carbon shaming," and with shamers such as environmental activists, environmentally conscious individuals, NGOs, the media, and intergovernmental bodies, rather than government regulators and administrative agencies. Climate shaming should be differentiated from the more general term of "eco-shaming," which relates to shaming in response to various types of activities that are considered harmful to the environment.

As is evident from the discussion of climate-change regulation in the previous section, climate change is a complicated topic. Consequentially, climate shaming is not an easy task, especially when it takes as its audience the general public and not professionals. This is a major challenge for climate shamers, who in order to be effective need to be able to communicate their message clearly and persuasively. For example, they need to explain succinctly how certain industrial or consumer activities are bad for the environment. As the causal link between the shamed behavior and climate change becomes less immediate and

---

[7] See, for example, the Sabin Center for Climate Change Law's Climate Reregulation Tracker, https://climate.law.columbia.edu/content/climate-reregulation-tracker.

obvious – as in the case of financial investments in carbon-intensive sectors, for example – climate shaming becomes more challenging.

Climate shaming can be carried out in various ways, all of which ultimately publicly highlight a socially undesirable behavior, with the aim of provoking feelings of shame in those who are considered as contributing to climate change, or of provoking sufficient public outrage to force them to change their ways. Whatever the chosen mechanism, climate shaming is fundamentally based on the anthropogenic characteristics of climate change (Aaltola, 2021) – that is, since climate change is caused by human actions (IPCC, 2021), people can be held morally responsible for their contribution to climate change.

While levels of condemnation of climate-related behaviors may vary from mild to harsh, public expressions of condemnation generally signal that an important value has been harmed (Lamb, 2003) – in this case, our safety, our health, our well-being, our very future. In this regard, successful climate shaming is perhaps less challenging a prospect because it focuses on a natural and obvious moral cause. Of course, there are still those who question the science behind climate change, expressing doubt as to whether climate change is attributable to human actions or even exists, but these opinions are becoming less and less dominant (Bell et al., 2021; Leiserowitz et al., 2021a).

Another core problem with climate shaming is that the public may care more about local pollution they can observe and feel the effects of, like sewage in a river or smog, than about more distant pollution that contributes to climate change on a global level (Ansolabehere & Konisky, 2014; Cohen & Viscusi, 2012; Downar et al., 2021). In addition, the gradual rate of escalation of climate change makes it difficult to effectively communicate information about the threat it poses (Teichman & Zamir, 2022).

However, recent research points to an increase in people's concern about climate change after experiencing extreme weather events (Hughes et al., 2020; Konisky et al., 2016; Leiserowitz et al., 2019), which unfortunately are now becoming more and more frequent (WMO, 2021a). In this vein, a recent Pew Center survey of 16,000 people in seventeen countries found that the majority of respondents, especially young adults, are now greatly concerned about climate change (Bell et al., 2021). According to the survey, most people are worried that they will suffer from the effects of climate change during their lifetimes and are willing to take personal steps, such as lifestyle changes, in response. Another international Pew Center survey, from 2018, found that majorities in most countries perceive climate change as a major threat to their country and as the greatest international threat today (Poushter & Huang, 2019).

Certainly, climate change has received greater public attention and recognition in recent years and has become the subject of much rightful concern.

Accordingly, people are now more open to government implementation of various climate policy and regulatory tools (Bergquist et al., 2020; Leiserowitz et al., 2021c). They are also engaging in climate actions such as demonstrations, marches, strikes, consumer boycotts, public expressions of criticism and disapproval, and personal behavioral changes (IPCC, 2022b: ch. 5; Leiserowitz et al., 2021b). The COVID-19 pandemic may have also contributed to our understanding that "invisible threats" can give birth to very real global health and environmental crises with very real impact on our lives (Geiger et al., 2021).

In recent years, shaming has become an increasingly prominent element of social and political efforts to mitigate climate change. For example, the international community harnesses shaming to pressure states to commit to and achieve ambitious reduction goals for greenhouse gas emissions (Spektor et al., 2022). A case in point is the Paris Agreement, which is largely based on negative reputational consequences for countries that fail to fulfill their pledges (Jacquet & Jamieson, 2016; Lyster, 2021). Under the Agreement, countries report their progress and other countries, as well as local and global public opinion, hold them accountable (Tingley & Tomz, 2022).

NGOs, too, contribute to the climate shaming of nations, for example, by producing rankings of countries based on their pledges, energy use, climate policy, and greenhouse gas emissions. For instance, the Climate Action Tracker rates governments' climate policy responses in categories ranging from "critically insufficient" to "almost sufficient."[8] Similarly, the World Resources Institute presents all countries' nationally determined contributions (NDCs) to the reduction of greenhouse gas emissions in accordance with the Paris Agreement on an interactive map, highlighting countries that have only submitted initial, rather than new or updated, NDCs (Fransen, 2021). Another organization publishes the Climate Change Performance Index, which labels countries as "winners" or "losers" based on their climate policies and achievements (Burck et al., 2021). Similar publications and rankings are offered by various media outlets.[9]

Individual activists also work at shaming countries into better climate law, regulation, and policy. Notably, Swedish activist Greta Thunberg, who is considered by many as a climate-change icon, is well known for her shaming tactics directed at world leaders, especially surrounding COP meetings, when she calls out political leaders' passivism and charlatanism in connection with climate policies (Aaltola, 2021).

---

[8] https://climateactiontracker.org/countries.

[9] See, for example, the *Financial Times*'s ranking of states' emissions and pledges, www.ft.com/content/9dfb0201-ef77-4c05-93cd-1e277c7017cf.

Additionally, both individual activists and NGOs create a shaming effect via climate litigation, in which public attention is drawn toward countries that, for example, fail to legislate or implement climate laws or submit insufficient NDCs.[10] Such litigation can signal the moral flaws of the defendant, and a breach not only of a legal but also of a social norm (Carlarne, 2021; Haines & Parker, 2017; Shapiro, 2020).

Climate shaming is also happening on an individual, social level, as people attempt to shame others for their carbon footprint – that is, for performing various everyday activities that indirectly contribute to climate change, such as shopping, heating, driving, and flying. For example, one of the most familiar and arguably effective climate campaigns – launched by the then fifteen-year-old Thunberg – has prompted a phenomenon known as "flight shaming," in which people, especially public figures, are publicly disgraced for their contribution to the global carbon emissions problem through taking flights (Mkono & Hughes, 2020).

However, the climate shaming of nations, as well as individuals, remains limited in many respects. Despite NGO efforts to "track and shame" nations, international efforts to create shaming mechanisms that will nudge countries to do better, climate litigation against countries, and Thunberg's persistent shaming of world leaders, the world is still not on track to meet the 1.5 °C goal.

The effectiveness of shaming individuals is also questionable, as each individual's contribution to climate change through various everyday activities is extremely small in comparison to the fossil-fuel companies known as "carbon majors" (Jacquet, 2015). In fact, more than two-thirds of all greenhouse gas emissions are attributed to some 100 such carbon majors worldwide (Heede, 2014, 2020). To illustrate this point, a recent report has found that the annual total of greenhouse gas emissions produced by Australia's leading carbon major is equivalent to the estimated emissions of twenty-five million Australians for the same period (Moss & Fraser, 2019). The report further indicated that Australia's six carbon majors together emitted five times more CO_2 in 2018 than all domestic transportation in Australia.

Climate shaming between individuals is also the least accurate type of climate shaming, as it often relies on rumors, speculations, anonymous reports, and information taken out of context. Finally, as will be discussed in further detail in Section 1.3, shaming individuals, by any type of agent, arguably carries far greater moral jeopardy than the shaming of other kinds of targets, such as artificial entities (Jacquet, 2015; Nussbaum, 2004; Yadin, 2019a).

---

[10] See the Sabin Center for Climate Change Law's Climate Change Litigation Database, http://climatecasechart.com.

1.3 Shaming

The concept of shaming carries different meanings across various contexts and disciplines and has no single clear definition. According to Gee & Copeland (2022), for example, shaming is the action of expressing condemnation of a characteristic or behavior to an audience, with the intention of invoking a shame response and a change in behavior consistent with the shamer's perceived norms. Van Erp (2021) defines shaming as the public condemnation of a person or corporation, extending to its exclusion from social networks, loss of reputation, and loss of opportunities. And John Braithwaite's (1989) highly cited definition of shaming refers to "all social processes of expressing disapproval which have the intention or effect of invoking remorse in the person being shamed and/or condemnation by others who become aware of the shaming."

A further examination of the concept of shaming reveals different points of view as to whether shaming is an internal or external process. Some researchers, for example, stress the act of shaming and the ways in which it is perceived and carried out by shamers (Jacquet, 2015; Lamb, 2003). This approach to shaming is external, focusing on the process of "private enforcement" by individuals and organizations who generate a morally negative response via, for example, denunciation, ostracism, disapproval, disrespect, harsh criticism, or condemnation (Lamb, 2003). The shaming of public and private organizations is mostly concerned with the external shaming approach.

Other researchers focus on the person who is being shamed and the inner processes that take place within that person's mind (Fredericks, 2021). This approach to shaming (the internal approach) predicates that shaming is dependent on feelings of shame, rather than on the acts or feelings of others toward the shamed person. It should be noted that shame itself is commonly defined as a negative self-valuation, accompanied by self-awareness of the ways in which one's faulty personality may be reflected to others (Tangney et al., 1995). Generally, internal and external aspects of shaming may coexist, but they can also take place independently of one another.

Among various shaming actors and targets, the most popular and prominent form of shaming today is probably the shaming of individuals by other individuals, especially on the Internet. People shame others for all kinds of behaviors and attributes, not just those related to climate change (discussed in Section 1.2). Examples include parking in spaces reserved for people with disabilities; carrying out sexual harassment or abuse (as highlighted by the #MeToo movement); being overweight ("fat shaming"); or not wearing protective masks ("pandemic shaming") (Gee & Copeland, 2022). Notice that this type of shaming is not dependent on any formal legal process. Rather, it is used as a kind of "social justice" tool, directed at a person considered to have acted illegally, immorally, or otherwise inappropriately, at least in the eyes of the shamer (Solove, 2007). Shaming can therefore occur in conjunction with, before, after, as a result of, or independently of formal legal action, such as criminal, administrative, or civil proceedings.

The growth of social media networks and other online platforms has resulted in a substantial increase in the scope, scale, and potential impact of shaming activities (Klonick, 2016). In the Internet age, it is all too easy for people to repeat the initial act of shaming and to publicize the humiliating information themselves, as well as to shun, ridicule, undermine, and emotionally or financially hurt the person being shamed. In these and other cases, shaming can result in an emotional harm so great that it causes the shamed individual the equivalent of physical pain and may never heal (Williams, 2007).

Against this background, it is not surprising that shaming is often regarded as the modern, technological form of stoning and lynch-mob justice (Whitman, 1998). This perspective considers shaming to be immoral, undemocratic, and disproportionate; a despicable form of action that needs to be eradicated (Nussbaum, 2004; Solove, 2007). A small number of studies also discuss the emotional and moral damage that shaming may hold for the shamer (Fredericks, 2021), and the costs that shamers incur from limiting their social and economic interaction with the shaming targets (Skeel, 2001). Others also underscore shaming's questionable efficacy, explaining that since public norms change over time and vary by location and communities, shaming is a complex action with uncertain results (Gee & Copeland, 2022; Van Erp, 2021).

A prominent strand of shaming scholarship, especially within law and criminology, discusses shaming that is carried out within the legal system by formal legal institutions, most notably by criminal courts. In fact, the history of criminal law is rooted in punishments such as public whipping, searing the mark of Cain on the forehead of the lawbreaker, or using pillories (Massaro, 1991). These sanctions included a component of public moral denunciation and were characterized by a purposeful direction of attention toward the criminal's act (Massaro, 1991). The goal was to reinforce pervading social norms and law-obeying culture by denouncing the nonconforming behavior of the shamed individual (Kahan, 1996).

Today, criminal courts sometimes include shaming as part of plea bargaining, as an alternative to traditional sanctioning such as incarceration, penalties, or license revocation (Garvey, 1998). In these cases, defendants may be required, for example, to publish a newspaper apology, to publicize previous drunk driving offences on their license plates, or even to carry a street sign detailing their wrongdoings (Garvey, 1998; Skeel, 2001). While many social science and legal scholars view this type of shaming as a harmful or ineffective practice that should be eradicated (Kahan, 2006; Massaro, 1991; Nussbaum, 2004), others believe that it can serve as an effective punishment that achieves worthy outcomes, maintains civil order, and deters undesirable behavior (Etzioni, 2003; Kahan, 1996; Whitman, 1998).

In this Element, I focus on another type of governmental shaming, executed by administrative regulators. As will be discussed in further detail in Section 2, I use "shaming" to refer to the action of drawing negative public attention to the behavior of a named or specific artificial entity or to a group of such entities, in a manner that may affect their behavior so that it better aligns with the public interest. I therefore utilize the external rather the internal approach to shaming. More specifically, I use shaming to refer only to the initial action of publicizing potentially damaging information, and not to subsequent shaming or other action by third parties, such as various stakeholders (though in its full form it does include such a component).

Additionally, the shaming discussed in this Element does not necessarily require reputational harm to occur; rather, the threat of such harm is designed to serve as a sufficient deterrent of unwanted behavior. "Shaming" as I refer to it in this Element is also not dependent on whether such governmental publications are effective or not in promoting the public interest. In other words, my working definition of "shaming" is focused on the behavioral (the action itself of publicizing negative information), rather than consequential dimensions of a governmental action (the result of this act of publication). Finally, while in theory such an action can cause reputational damage even without intention, I will mainly discuss intentional shaming.

From different perspectives, shaming carried out by the state (via courts or governments) can be viewed as either harsher or softer than private shaming between civilians. To give one example, shaming by the executive branch can be perceived by its audience as more credible and reliable because it originates in an authoritative body of government (Cortez, 2018). As a result, a governmental publication may have a greater shaming effect than civil shaming between individuals.

Public perception in this context, though, may vary based on the identity of the shamer. The public usually trusts regulatory administrative agencies more than elected representatives due to the credibility, rationality, and transparency afforded by administrative procedures (Stiglitz, 2018), and the stewardship and expertise that such agencies have (Bratspies, 2009). However, government entities may also be corrupt, and use shaming as a retaliation tool or to serve a private interest (Cortez, 2011); authoritarian and autocratic regimes may use it as a rights-infringing sanction; and authorities may also err, for example, by mistakenly publicizing "fake news." Under such circumstances, governmental shaming may lose credibility and effectiveness and therefore be regarded as a soft form of regulation.

Legal limitations imposed on shaming entities offer another point for discussion in this context. While regulatory agencies, for instance, are subject to constitutional and administrative review of their actions, individuals often shame anonymously and are therefore less accountable for their online actions. From this point of view, governmental shaming can be regarded as a softer action than private shaming.

1.4 Intended Contribution

This Element aims to contribute to the literature in several research fields, chief among them RS, climate-change regulation, and environmental disclosure. It uniquely lies at the intersection of these fields, offering an original theory and useful designs of RCS from regulation, law, policy, and governance perspectives. The Element is based on conceptual, theoretical, descriptive, normative, and policy-oriented analysis, informed by multidisciplinary study and an examination of prominent examples of RCS schemes in various jurisdictions.

2 Regulatory Climate Shaming – Conceptual and Theoretical Analysis

Modern regulatory strategies utilize an array of tools and approaches, and usually do not rely solely on the command-and-control approach, which involves legal prohibitions accompanied by criminal and administrative sanctioning (Ayres & Braithwaite, 1992). Variations and combinations of both command-and-control and softer approaches – such as self-regulation, economic incentives, voluntary regulation, contractual regulation, and disclosure regulation – are often utilized by regulators to tackle both old and new challenges in markets and in industry sectors (Baldwin et al., 2012). This is also true for the environmental (Gunningham & Sinclair, 2017) and climate regulatory landscape (see Section 1.1). This Element focuses on one intriguing such approach, known as regulation by shaming.

While RS is beginning to develop as a research field of its own, it is most closely related to disclosure regulation. The discussion therefore draws most prominently on research into RS and disclosure regulation. Environmental policy and climate-obstruction literatures are also referenced in this section.

In Section 2.1, I begin by exploring the conceptual, descriptive, theoretical, and normative dimensions of RS, reviewing and evaluating the extant theories and empirical evidence regarding the effectiveness, desirability, and legitimacy of RS in various fields, forms, and jurisdictions, especially in the environmental context. Then, in Section 2.2, I introduce a regulatory climate-shaming framework and put forward a normative theory for this approach, based on its unique rationales, mechanisms, advantages, justifications, and challenges.

2.1 Regulatory Shaming

Regulatory shaming refers to the practice by regulators in the executive branch of intentionally publishing details of corporate misdeeds in a manner that conveys a negative message to the public about misbehaving corporations, so as to encourage them to comply with mandatory norms and/or adopt voluntary norms, utilizing social pressure and corporate reputational sensitivities (Yadin, 2019a). Shamed entities are mostly specific, named companies, but can also include various other types of corporations, businesses, and nongovernmental organizations, as well as entire industries and sectors. Shaming actors may include governments, ministries, regulatory agencies, or municipalities – indeed, all levels of government can engage in RS. While largely focused on negative publications, RS mechanisms may also include the provision of positive information about other companies – for example, by highlighting good practices, as well as with ranking and scoring mechanisms in which some companies are graded low, while others are ranked high.

In general, there are several essential components to the RS process (Yadin, 2019c): (a) choosing a topic for RS that people will be interested in or passionate about, that is noncontroversial and relatable, and that can be easily understood by relevant stakeholders; (b) identifying the right stakeholders and the right shaming targets; and (c) designing shaming messages and techniques that are suitable for the relevant stakeholders and shaming targets, as well as the shaming goal.

Regulatory shaming typically aims to protect the public interest – such as public safety and health, consumer rights, competition in markets, and the environment. Like other types of regulation, it is aimed at correcting market failures, such as informational asymmetries and negative externalities, and advancing desired social goals, rights, interests, and values.

It is usually carried out through publications concerning illegal, inappropriate, or immoral corporate activities, as well as adverse corporate characteristics. Regulatory shaming can also focus on other aspects, such as business practices, performance in markets, or customer satisfaction. For example, the UK Financial Conduct Authority (FCA) publishes consumer complaints (which are not necessarily indicative of any legal infringement) together with the names of the firms.[11] Within this regulatory framework, firms that are able to significantly reduce the number of complaints are exempt from such shaming and are thereby motivated to improve performance.

More specifically, the publicized information may refer to compliance, non-compliance, or "above-compliance" pertaining to administrative, civil, criminal, and voluntary norms. For example, RS may focus on publicizing information about formal legal proceedings and their outcomes, such as the issuing of citations or imposition of civil penalties following a corporate violation of a legal norm. It may also publicize information on corporate compliance, in comparison to other firms as well as to the legally binding standard. For example, if a legal norm requires large companies to employ people with disabilities to the extent of 2 percent of the workforce, the information may detail companies' actual hiring, thereby exposing noncompliance and showcasing compliance and "above-compliance."

In cases where there is no violation of any legal norm, RS aims to nudge firms to comply with corporate social responsibility (CSR) norms, building mostly on companies' "social license." Under the terms of CSR, the corporate entity is understood through a communitarian prism, which focuses on social and moral aspects of the corporation's activities, rather than merely on its own self-interest (Branson, 2002). This approach has given rise to the "stakeholder model," in which shareholders are considered only one of the interest groups to which the corporation is beholden (Branson, 2002). Relatedly, "social license" governs the extent to which corporations are constrained to meet societal expectations and avoid activities that society deems unacceptable, whether or not these expectations are embodied in the law (Gunningham et al., 2004). In many ways, shaming based on firms' social license is a more basic form of RS, as it relies solely on eliciting public responses, while shaming based on noncompliance also relies on an accompanying legal sanctioning procedure (Yadin, 2019b). At the same time, it has been argued that shaming based purely on voluntary norms incorporates a broader understanding of regulatory agencies' modern roles and capabilities (Yadin, 2019b).

In the case of noncompliance with a legally binding norm, the goal is to impose multilayered costs on firms that exceed the damages they might incur as a result of traditional penalties or monetary fines, and thus, to better incentivize them to comply with regulatory norms. In cases of shaming based on voluntary norms, the goal is to create incentives for corporations to become more socially responsible regardless of the existence and the enforcement of laws, rules, or regulations. In both cases, firms are expected to accommodate their behavior

---

[11] www.fca.org.uk/data/complaints-data/firm-level.

based on actual or fear of reputational harm (Stephan, 2002). In fact, the mechanism of shaming in itself, or even the prospect of its implementation, is meant to motivate both compliance and "beyond-compliance."

Regulatory shaming rests on the most popular understanding of "regulation" as being any activity of the executive branch of government, performed by a national or local administrative authority, which aims to control or influence the behavior of nongovernmental organizations (such as corporations) that operate in markets and industry sectors, in order to protect the public interest (Koop & Lodge, 2017). Regulatory shaming further rests on the working definition of shaming discussed in Section 1.3, which focuses on the action of drawing negative public attention to the behavior of a named or specific artificial entity or to a group of such entities. As corporations are artificial entities and shame is a human emotion, RS is instead founded on leveraging corporate sensitivity to reputational gains and losses, rather than on inflicting emotional harm.

Generally, RS can be based on information provided by regulatees themselves or gathered independently by regulators, or a combination of both. For example, it can utilize the output of companies' own reporting obligations, of regulatory on-site inspections, or a mixture of both. Information can also be gathered by the public, for example, through a governmental complaint database, constituting a form of crowdsourced monitoring (Yadin, forthcoming-c).

The information made public may be detailed or summarized, raw or processed, technical or substantive; and it can take many forms, including star, grade, or color ratings, league tables, public statements, online databases, labeling schemes, and publication of enforcement actions or inspection results (Fung et al., 2007; Yadin, 2019a). These are disseminated through both old and new media, including websites and apps, social media, press releases, product labels, newspaper ads, signs in places of business, and corporate announcements and reports.

As discussed in Section 1.3, shaming methods have been massively enhanced in the digital age, in which sophisticated yet very accessible, low-cost, and simple-to-operate online platforms enable shamers to reach large audiences in a matter of seconds, and also to target relevant audiences (Meijer & Homburg, 2009). The same platforms are also harnessed by administrative regulators, who can now also easily and quickly disseminate shaming information to a large number of people, who can digitally (or nondigitally) continue the shaming process. Regulators can widely share, for instance, consumer complaint databases, digital maps of violations, or various announcements, documents, videos, pictures, and infographics (Cortez, 2018). Regulatory shaming publications are now searchable, downloadable, and interactive, allowing users to view and compare data on companies in various formats. Digital platforms also allow regulators to continuously update the data they post online, while maintaining direct and constant communication with the public on corporate performance and behavior. It has been argued that in this way, RS fosters new forms of communication and relationships between government, citizens, and corporations (Yadin, 2020).

Regulatory shaming should be differentiated from other types of expressive regulatory actions, since it is often presented, regarded, or misunderstood as mere disclosure or transparency (Fung & O'Rourke, 2000; Van Erp, 2010; Yadin, 2019c). Indeed, various regulatory information-sharing schemes have mixed goals, framings, and impacts, which may cause some conceptual overlaps and confusion. These may include informing (supporting decision-making, for example of consumers or investors), educating, warning, nudging (influencing individual choice and affecting behavioral change), promoting legal and regulatory certainty (e.g. by publicizing enforcement information), protecting the public's right to know (for instance, about hazardous corporate activities), promoting governmental transparency, protecting people's personal autonomy (by giving them the ability to choose in accordance with their preferences and not be misled), or shaming (conveying a negative message on corporate behavior to encourage compliance).

In terms of volume, only a small (though growing) portion of governmental information sharing is shaming. Additionally, shaming publications often promote other goals of information sharing, and are rarely restricted just to shaming. Thus, some forms of regulatory publications have a greater element of shaming than others. For example, while the main purpose of labels and similar disclosure schemes is usually to support decision-making, and/or to warn, educate, and nudge individuals, they may also carry implicit shaming messages. Companies that are forced to label their products as unhealthy, for instance, may be shamed in the sense that this information can potentially damage their reputation and invite relevant shaming communities to apply pressure to alter their business models. These concepts are illustrated in **Figure 1**.

Regulatory shaming therefore relates to a spectrum of information-sharing schemes, extending from soft forms of shaming at one end, such as certain labeling schemes, to harder forms of shaming at the other, such as condemnatory statements singling out specific companies. The ways in which the information is presented affect the level of shaming, depending, for example, on the wording of the message; the use of colors (such as red) and scores; the ranking methods used; the attachment of pictures (e.g. of noncompliant facilities); the use of social media or other media, which can be more or less visible and accessible; and choosing to focus on one firm or more (the shaming of multiple firms can be considered diffused and thus softer, especially in schemes such as databases containing thousands of firms).

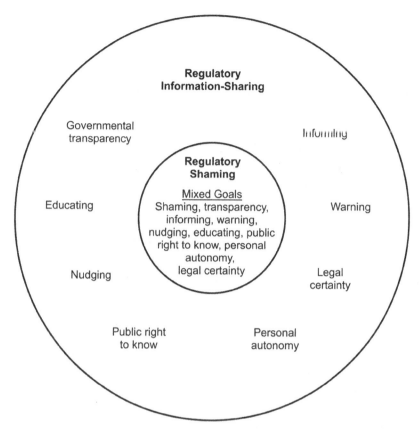

Figure 1 The different goals of regulatory information-sharing

Indeed, RS is closely related to, and often builds on, disclosure mandates. However, it should be differentiated from the tactic known as disclosure regulation. Generally, disclosure regulation requires companies to disclose information in order to help consumers, investors, users, viewers, patients, and other stakeholders to decide whether, how, when, where, and how much to use a product or a service (Ben-Shahar & Schneider, 2014; Fung et al., 2007). Yet while both disclosure regulation and RS tools are based primarily on making information publicly available, RS involves a negative judgment and the expression of normative disapproval by the regulator (see also Section 1.3). A shaming message may express a regulator's dissatisfaction, disapproval, scolding, or condemnation, and it will highlight the shamed entity's unacceptable behavior, character, or values and morals.

To demonstrate the different types of regulatory information sharing, it is worth considering food content regulation. While health regulators require food companies to disclose calorific values on packaged foods, the information in itself does not carry a message of negative judgment by the regulator, but merely aims to inform consumers so that they can make a conscious, facts-based choice. By contrast, regulatory schemes that rank food manufacturers according to the levels of trans fat, sugar, and salt in their products, possibly accompanied with condemnatory regulatory statements, can be considered highly shaming. In between these models lies a burgeoning regulatory system of front-of-package warning labels. Such labeling systems typically include the use of colors (within a "traffic-light" system) and icons to visually signal to consumers which foods are unhealthy and should be avoided or consumed very moderately (Yadin, 2021b). This regulatory mechanism combines informational support for consumer decision-making, warnings to consumers, promoting the public right to know and citizens' personal autonomy, educating, and shaming of food companies that market unhealthy foods. Such shaming aims to induce firms to modify their product ingredients, introduce new products, and eliminate old ones so that fewer warning labels, which may damage their sales and reputation, are placed on their products.

Generally, RS invites relevant audiences to alter their behavior, discourse, or ways of thinking with regard to the shamed entity and to engage in disapproval, criticism, condemnation, protest, excommunication, boycott, or social, legal, and political activism of various kinds. Examples are provided in **Figure 2**. Stakeholders may share or publicize the original shaming publication (e.g. via social media), or respond to it more actively (e.g. via consumer boycotts).

Regulatory shaming has gained momentum in recent years across different jurisdictions and fields. In the field of public health, which is closely related to climate change, one prominent example of RS consists of the tweets and press releases routinely issued by the US Occupational Safety and Health Administration (OSHA) about occupational safety violations. In these publications, formally framed by the agency as a regulation-by-shaming policy, OSHA not only names specific companies but also condemns their behavior (Yadin, 2019b). Thus, publications may include statements regarding issued citations and settlement agreements, with an identification of a specific company, a detailed description of its worker safety violations, the implications for employees' health, and a moral judgment of the company's behavior. For example, one OSHA news release stated that a named company's "history of safety violations continues, putting employees … at risk of serious injuries," and that the company's "extensive list of violations reflects a workplace that does not prioritize worker safety and health."[12]

Another example can be found in US pharmaceuticals regulation. The US Food and Drug administration (FDA) recently published a list of pharmaceutical

---

[12] www.osha.gov/news/newsreleases/region2/07212017.

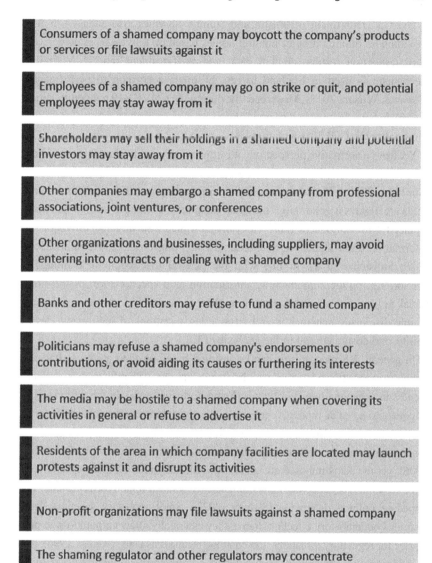

Consumers of a shamed company may boycott the company's products or services or file lawsuits against it

Employees of a shamed company may go on strike or quit, and potential employees may stay away from it

Shareholders may sell their holdings in a shamed company and potential investors may stay away from it

Other companies may embargo a shamed company from professional associations, joint ventures, or conferences

Other organizations and businesses, including suppliers, may avoid entering into contracts or dealing with a shamed company

Banks and other creditors may refuse to fund a shamed company

Politicians may refuse a shamed company's endorsements or contributions, or avoid aiding its causes or furthering its interests

The media may be hostile to a shamed company when covering its activities in general or refuse to advertise it

Residents of the area in which company facilities are located may launch protests against it and disrupt its activities

Non-profit organizations may file lawsuits against a shamed company

The shaming regulator and other regulators may concentrate rulemaking, monitoring, and enforcement resources on the company

Figure 2 The regulatory shaming mechanism: stakeholders' responses

companies that the agency claims act unethically, and possibly unlawfully, in the markets.[13] This "shaming list," which was posted on the FDA's website, includes the names of branded drug companies that allegedly tried to block competition from generic drug companies (Yadin, 2019c). According to the FDA's statement that accompanied the list, these companies are suspected of "gaming the system" in such a way that drives up drug prices.

---

[13] https://perma.cc/7XHY-NQDT.

The US Department of Health and Human Services also engages in RS, providing an online rating of nursing homes based on a five-star scale derived from inspection results, clinical data, and staff–resident ratio.[14] A similar mechanism is used in England to rate hospitals according to performance (Bevan & Wilson, 2013). Most recently, during the COVID-19 pandemic, cities in Canada have "named and shamed" businesses via website postings for failing to comply with COVID regulations.[15]

Yet from a normative perspective, it is now widely established that providing the public with more information on corporate activities is not always beneficial (Sunstein, 2020). The idea that informational schemes of various forms and goals are always a good thing, based on the public's "right to know," has been replaced with notions of more carefully designed and targeted provision of information (Ben-Shahar & Schneider, 2014; Fung et al., 2007; Shimshack, 2020; Sunstein, 2020). In line with the general discussion of shaming in Section 1.3, publicizing shaming information as a specific form of informational policy is considered more complex, delicate, and controversial than simple disclosure schemes, thereby rendering questions of legitimacy, desirability, and design central to the analysis of RS.

In general, RS mechanisms can be justified based on their ability to balance fairness toward affected parties, mainly shaming targets, with efficiency and effectiveness in achieving regulatory goals (Yadin, forthcoming-a). In this context, it is important to note that modern regulation is greatly lacking in efficient and effective enforcement tools. Notably, as corporations cannot be incarcerated, monetary sanctions remain the most commonly used corporate enforcement tool (Ainslie, 2006), yet the sums imposed are often low, especially relative to the revenues of large corporations, rendering noncompliance an efficient option for firms and thus impeding optimal deterrence (Shapira, 2022). Since monetary sanctions lack a strong condemnatory effect or stigma, they essentially allow corporations to pay a price for regulatory violations and continue business as usual (Kahan, 1996).

Additionally, since both criminal and administrative sanctioning require considerable regulatory resources, and are lengthy processes with uncertain outcomes, enforcement costs often outweigh the benefits. Regulatory enforcement of this type depends on establishing an extensive factual basis for sanctioning through rigorous investigations, inspections, reviews, and judicial procedures. For various legal reasons, these intensive efforts may still result in exoneration or in the revocation of sanctions after lengthy hearings in courts and other judicial tribunals. As a result, regulators often decide against pursuing

---

[14] www.medicare.gov/nursinghomecompare/search.html.

[15] See, for example, https://web.archive.org/web/20220319033624/https://www.hamilton.ca/cor onavirus/faq-about-enforcement.

formal enforcement actions and are left, in effect, with no means for creating optimal deterrence.

Furthermore, regulators are often unable to impose such hard-law sanctions due to political, legal, budgetary, or various other constraints – that is, they do not always enjoy the necessary political support, legal frameworks, or funds to take criminal or administrative action. Moreover, regulators are not always well positioned to promote legislation of command and control mechanisms of the type that would set legally binding standards and equip the regulators with appropriate hard-law enforcement tools, including necessary powers and resources. This type of legislation process also tends to be lengthy and complex. Consequently, the regulatory enforcement world is very much in search of new, more accessible, more efficient, and more effective methods for increasing corporate deterrence and compliance. ·

Regulatory shaming is arguably a much cheaper and speedier option than command-and-control sanctioning, litigation, monitoring, inspection, and legislation. This is mostly because shaming involves communication – conveying information, beliefs, and ideas mainly through digital media channels, which are virtually costless. In some cases, it is the corporations themselves that finance the shaming, such as when they are required to place signs and apply labels to goods and services, provide financial reporting, or publish apologies in the media, all of which are sometimes mandated by regulators as a form of shaming. Though the compilation and analysis of the relevant data by the regulator – which may include creating rankings, league tables, or searchable databases – may entail some expenses, these are still relatively small. Additionally, since shaming allows regulators to quickly publicize information on corporate performance, compliance, or ethics, and effectively reach a large and relevant crowd that can operate the shaming mechanism, it can make regulators more responsive to various threats. This advantage of immediacy may also be especially useful in times of crisis and emergencies.

Regulation by shaming can also induce corporations to adopt CSR norms. This can be useful in cases where command-and-control is not available for regulators, as discussed above, but also in other situations, as a supplementary tool to binding legal obligations. When command-and-control is indeed available, shaming can enrich regulators' "enforcement pyramids," which according to Ayres and Braithwaite (1992) should also include varied mid-level enforcement tools for optimal deterrence and compliance. Shaming can also function as an experimental tool to be used before proceeding to more formal modes of regulation, in cases where new or urgent challenges arise involving new types of firms, problems, and social and economic activities.

Additionally, RS is a form of private enforcement that is carried out by private individuals and organizations without formal contractual mechanisms of governmental compensation. Since RS often involves elements of crowd-sourcing – as it rests on the notion of many individuals sanctioning a firm – it can spread the costs of shaming without overburdening each shamer (Yadin, forthcoming-c). Thus, instead of leaving entire forms of business activities under-regulated or even unregulated due to the limitations of command-and-control apparatus, regulators can use the resources of various stakeholders to carry out regulatory enforcement.

However, RS is not merely an answer to the problem of shortage of regulatory resources in government. It can also open new avenues for regulation and enable new regulatory capabilities. Namely, RS does not merely transfer regulatory functions of administrative agencies to private entities for reasons of efficiency, but harnesses the public to perform regulatory roles that cannot be performed by agency staff. For example, consumer boycotts or mass withdrawal of financial investments in capital markets are steps that can only be taken effectively by the public in large numbers, and not by any regulatory agency. Relying on the public for regulatory tasks entails other types of benefits as well. Providing people with the opportunity to take part in the process of business regulation, for instance, can offer members of the public a healthy and productive way to channel their frustration and disappointment with underperforming corporations.

In addition, RS may be viewed as more democratic than other modes of governmental regulation, such as command-and-control, because it does not involve the government flexing its enforcement muscles (Yadin, 2019a). Instead, RS advances public participation in policy and promotes cooperation and trust between governmental and civic organizations, as well as individuals. According to this view, this process enables the government and its citizens to become partners in the endeavor to fulfill public goals and overcome social and economic failures. This is especially important in an era in which citizens' trust in government, including its bureaucratic and regulatory systems, is diminishing (Devine et al., 2020). Indeed, the recent COVID-19 pandemic has demonstrated the importance of citizens' trust in governments in times of global catastrophes and scientific uncertainty (Fancourt et al., 2020).

In a similar vein, RS can enjoy legitimacy because it can be regarded as a soft form of regulation. Though the regulator creates the conditions for shaming and initiates the shaming process, it does so with a light touch, a nudge, rather than being involved in the markets directly. Moreover, in this way, justice is delivered directly by the crowd and not by (usually) unelected officials. Regulatory shaming can similarly be regarded as a voluntary, soft-law tool, as firms can choose whether to respond to the shaming process and there are normally no legal sanctions on corporations for ignoring shaming publications. These characteristics of RS may also diminish regulatory confrontations between government and industries.

In some cases, RS may also present a more legally feasible policy option than command-and-control. In cases where authorizing legislation is needed, the fact that shaming schemes are often regarded a softer form of regulatory intervention in markets means that they can be more easily legislated. In other cases, RS may arguably be based on broad, nonexplicit statutory mandates, sometimes accompanied by agency procedures or rules. This is because it does not involve classic criminal or administrative sanctioning of firms, which generally require explicit statutory authorization.

The conceptualization of shaming as a soft form of regulation may also provide opportunities to create legal frameworks for RS based on consensus with relevant industries. Such a consensus may be reached against the threat of command-and-control legislation and the promise that RS can actually serve businesses. Namely, RS schemes can provide companies information about competitors, help improve business practices in a way that reduces the exposure of firms to various risks (Fung et al., 2007; Stephan, 2002), and also showcase firms that perform well, providing them with recognition and rewarding them reputationally. It should be noted in this context that faming frameworks may be legally and politically easier to legislate and implement than shaming frameworks, due to their less "sanctioning" character and perceived harms by industry members.

Importantly, empirical studies have also shown RS to be effective in various fields and jurisdictions, most notably in the field of public health, covering topics such as occupational health, food safety, and quality of health services. Regulatory shaming has also been proven to be effective in promoting various environmental goals relating, for example, to industrial air and water pollution and to environmental nuisances.

Studies have pointed to the effectiveness of various RS schemes in these fields. For example, RS based on star ratings has proven to be effective in the United Kingdom in reducing hospital waiting times and improving hospital performance (Bevan & Hood, 2006; Bevan & Wilson, 2013). Research has further demonstrated the effectiveness of health inspection grading posted on restaurant windows in the United States (via letters) and in Germany (via smileys) (Bavorova et al., 2017; Jin & Leslie, 2003; Simon et al., 2005). These studies have shown that these schemes improved compliance with food and hygiene regulations and resulted in fewer foodborne illness hospitalizations. Similar on-site publications of international (low) prices presented to consumers next to the local (high) prices of toiletry products in Israel have also led to some 8 percent decrease in prices (Ater & Avishay-Rizi, 2022).

Christensen et al. (2017) spotlight the feasibility of RS using mandatory reporting requirements for publicly traded companies. Their research studied the effect of a new regulatory requirement to disclose mine-safety records in Securities and Exchange Commission (SEC)-registered firms' financial reports, introduced following a 2010 disaster in West Virginia in which twenty-nine miners were killed. They found that this new disclosure requirement led to an 11 percent decrease in mining-related citations and a 13 percent decrease in injuries.

Press releases have also proven to be an effective form of RS in some cases. Johnson (2020) found that OSHA's press releases shaming companies for their violations, for example, has led other companies in the same sector or geographical area as the shamed entity to improve their compliance, resulting in fewer occupational injuries. According to the study, a single OSHA press release is equivalent, in terms of improvement in compliance, to more than 200 inspections. Huang et al. (2022) have found that nonpeer facilities located in other regions were also affected by the publication of these shaming messages and increased their safety measures by hiring more safety-related employees. In other words, RS can be used to reduce the required scale of classic regulatory monitoring and enforcement while maintaining the same levels of deterrence. Regulatory shaming can therefore, *inter alia*, improve compliance in under-regulated areas and by resource- and authority-limited regulators.

In the field of environmental protection, theoretical, qualitative, and quantitative studies in various disciplines, relating to a variety of industries in multiple jurisdictions, point to the effectiveness, feasibility, rationales, and mechanisms of information-based policies that include a corporate shaming component (which I refer to as "regulatory eco-shaming.") Generally, research suggests that corporate environmental reputation is regarded as important by managers, and that managers are inclined to improve performance in order to protect it (Gunningham et al., 2004; Prakash & Potoski, 2006). In fact, environmental managers perceive neighboring communities and the public at large as one of the top three factors influencing corporate environmental performance (Doonan et al., 2002). Research indicates that reputational sanctions by stakeholders, such as members of the public and local communities, can motivate companies not only to comply with environmental regulations but also to go "beyond-compliance" and join "green clubs," by adopting voluntary environmental programs such as ISO 14001 (Gunningham et al., 2004; Prakash & Potoski, 2006).

Studies suggest that what drives corporations to adopt environmentally responsible behavior is not only their legal license to operate but also their social license (Gunningham et al., 2003). Another explanation is premised on organizational behavior approaches that describe corporate behavior as a response to what other competing and successful firms are doing. According to this idea, known as institutional isomorphism (DiMaggio & Powell, 1983), corporations may adopt certain behaviors, such as implementing voluntary environmental standards, not only for reasons of efficiency but also because of cultural environment constructs (Gunningham et al., 2003: 32).

Corporate environmental reputation has been identified as a driver for improving compliance in a range of policy schemes that utilize a mixture of different stakeholders. For example, research points to the benefits of harnessing firms' sensitivity to their environmental reputations by creating and publicizing regulatory environmental rankings of companies. A study by the World Bank focused on a government program in Indonesia, which assigns color ratings to factories based on their performance (Afsah et al., 1996). The scheme included five colors, with a gold rating awarded to factories that achieved above-compliance standards, and a black rating given to factories that made no attempt to control pollution and were causing serious danger. Before the assigned ratings were released to the public, the companies were notified of their scores and were given time to improve them. During that period, half of the "black" plants succeeded in upgrading their status. The results indicate high levels of responsiveness by industrial facilities to environmental reputational sanctions under certain conditions. Similar programs in developing countries have also proven to be effective in encouraging compliance, in more recent studies (Darko-Mensah & Okereke, 2013).

Similarly, a Dutch study focused on the effectiveness of a naming-and-shaming list, published by environmental agencies, of the top-ten companies that were most complained about by local residents (Meijer, 2013). The study, which was based on interviews of industry stakeholders, showed that the publication of the shaming list had motivated managers to get off it by reducing environmental nuisances. According to the study, companies that reacted strongly to the publication of the information understood it as harmful to their reputations.

Corporate reporting is another form of regulatory eco-shaming that has proven beneficial. Bennear and Olmstead (2008), for example, found that the mandatory disclosure of violations of drinking water regulations and containment levels to consumers resulted in a 30 percent to 44 percent reduction in violations, and a reduction in severe health violations of up to 57 percent.

The authors posited that this effect can be explained by water suppliers fearing that people will lobby for more stringent regulation of the field.

In addition, studies on the mechanisms and functions of environmental databases have pointed to their ability to incentivize firms to improve performance, fearing the reputational costs of negative publicity and public backlash (Cortez, 2018: 43–44; Fung & O'Rourke, 2000; Stephan, 2002). Fung and O'Rourke (2000), for instance, focus on the EPA's open public registry of factories' chemical pollution, known as the Toxic Release Inventory (TRI). This facility-based and firm-based online database on toxic chemical releases, established in the mid-1980s, is considered a flagship of environmental disclosure policies. Following the introduction of the TRI, reported releases dropped by more than 50 percent in a period of ten years (Fung et al., 2007). Since these reductions were not required by law, it has been suggested that firms have reduced their toxic releases due to public pressure, especially in response to blacklists and top-ten lists that the media and environmental organizations have periodically circulated based on the database (Fung & O'Rourke, 2000).

Indeed, these lists have generated actions and negative responses from various stakeholders, including citizens, journalists, investors, employees, and policymakers (Fung & O'Rourke, 2000). Similarly, in a more recent example Bonetti et al. (2023) show that mandated disclosure via public registries in the field of hydraulic fracturing ("fracking") has been successful in promoting environmental goals based on shaming carried out by NGOs, shareholders, and local newspapers.

Despite the benefits, advantages, effectiveness, and theoretical justifications of RS (in general and in the environmental context in particular), it is not devoid of costs, challenges, or risks. For example, RS entails administrative costs relating to time and effort in creating, legislating (in some cases), and implementing policies. During these stages of regulation, companies may also try to "capture" regulators and other policymakers in order to influence RS schemes in their favor, and to reduce their possible exposure to public and political pressure to improve performance (Fung et al., 2007; Fung & O'Rourke, 2000). However, as discussed in this section, the direct costs of constituting and implementing shaming schemes are generally low relative to other forms of regulation such as command-and-control. Regulatory capture is also not a concern unique to RS, and in fact, it may arguably be less serious in cases of soft regulation (such as shaming) than hard regulation, which is often perceived by industries as more threatening. Moreover, the crowdsourced nature of RS has the potential to reduce capture, especially in the policy implementation stages, because it is performed by a large number of diverse stakeholders (Yadin, forthcoming-c). This potential is backed by regulatory capture theory, which posits that it is harder for industries to capture and influence regulation and regulators when policymaking and enforcement is diffused (Laffont & Martimort, 1999).

Another challenge relates to cases in which RS policies are implemented, but companies manipulate their mechanisms and effects. For example, companies may try to manipulate the shaming process by anonymously engaging in the shaming of competitors on social media and other platforms. Firms may also react to shaming policies by narrowly focusing on adjustments that would improve their rank, grades, and scores, rather than making real improvements (Bevan & Hood, 2006; Fung et al., 2007: 72; Shimshack, 2020). This type of corporate behavior is especially relevant in complicated areas of regulation, where it is easier for corporations to manipulate the data covered by the shaming scheme (Wilson, 2004).

More recent research in the fields of public health and environmental protection also suggests that RS is subject to corporate manipulation. For example, Huang et al. (2022) suggest that following the implementation of OSHA's regulation-by-shaming policy, firms have improved safety performance by reallocating resources from areas such as environmental protection and financial reporting, which were not as publicly exposed to shaming as occupational safety infringements. It has also been argued that companies reporting chemical releases under the TRI framework have substituted chemicals that were subject to disclosure with hazardous chemicals not subject to disclosure, or have simply moved their facility outside the United States to avoid disclosure (Cohen & Viscusi, 2012).

Another type of corporate manipulation in this context is "creative compliance," in which regulated firms exploit legal loopholes in a manner that is perhaps not a violation of the law *per se*, but can neither be regarded as full compliance with the law. For example, Israeli food companies have implemented graphical manipulation tactics to impede a recent food-labeling policy based on round red warning markings, by changing the background colors of packages to red, among others (Yadin, 2021b). While the companies met the regulatory requirements in the technical sense, the way they complied was entirely against the spirit of the reform.

Other challenges of RS relate to the nature and consequences of the regulatory act itself. For example, RS schemes may be regarded as an abuse of power, unduly infringing on legitimate corporate interests and rights. Namely, they may violate corporate rights to procedural due process upon state deprivation of property, the right to a good name, and the right not to be subject to unauthorized administrative action. Regulatory shaming may also be disproportionally applied, exposing companies to the risk of extensive losses in a relatively short period of time with no real options for repairing their reputation.

This may be especially true for small businesses that lack the resources to effectively respond to the shaming.

Indeed, shaming is often unpredictable in its magnitude and effects, which may grow out of all proportion to the original misdeed. This is partly because the shaming sanction is so easily applied, especially in the age of social media and mass media platforms (see Section 1.3), and encourages public responses that can quickly spiral out of control. In this vein, RS allows the unfiltered emotions of stakeholders, such as consumer outrage and desire for revenge, as well as stakeholders' belief systems and individual inclinations, to be part of the regulatory process. Additionally, private persons do not always see "the big picture" or possess the required legal, regulatory, or other professional experience and knowledge to appropriately enforce regulatory norms. Regulatory shaming may therefore be regarded as unfair "mob justice" and create legal and regulatory uncertainty for industries, thus driving businesses away and suppressing growth and innovation.

In the same vein, shaming also involves low levels of accountability for stakeholders taking part in the shaming process, who may be inclined to exaggerate company misdeeds, overstate minor incidents, and use harsh language. Furthermore, people may be inclined to over-shame as part of a crowd mentality. Crowd dynamics may also fuel adverse feelings of anger, disappointment, and discontent with business entities, and thereby lead to an overall decrease in people's joy, fulfillment, and happiness in life. From a broader perspective, RS may foster an abusive shame culture that extends to the shaming of individuals, which is highly controversial (see Section 1.3). It could therefore be argued that RS is not a form of action that is appropriate for the state to take.

Additionally, while RS of artificial entities like firms may be regarded as legitimate since it does not affect its targets psychologically or emotionally (Yadin, 2019a), it may still result in reputational damage spillover to individuals like corporate officers and shareholders (Bevan & Wilson, 2013; Van Erp, 2011). Also, some point to a collective sense of shame that can be experienced by a group of people who belong to a shamed organization, such as employees (Fredericks, 2021).

Shaming fostered by the state may also prove inefficient, ineffective, and produce counterproductive results in other ways. For example, regulators who engage in shaming may find themselves being scolded or attacked by targeted companies or third parties. In these cases, regulators may harm their relationship with the industry, jeopardize their own reputations, and become entangled in costly and prolonged legal battles (Van Erp, 2007). As a result, RS may diminish public trust in government regulators. Shaming can also strengthen adverse corporate behavior, because it publicizes to all ill-behaved companies that their inadequate behavior is not uncommon, and is even standard. This may also produce a contagion effect of noncompliance among complying companies.

Regulatory shaming may also prove to be ineffective due to inappropriate design, in terms of the selection of topic, information, shaming targets, relevant stakeholders, and form of media. For example, the FDA's shaming list of branded drug companies was extremely uncommunicative in terms of both the language used and the ways in which the data was processed, organized, and presented; furthermore, it was not distributed through appropriate channels for effective impact (Yadin, 2019c). In another case, the Dutch financial regulator's policy of publishing all the names and violations of corporations and the sanctions imposed was found to be unsuccessful in effecting shaming and generating deterrence, because the information shared pertained to technical violations, and thus the inherent moral message was weak (Van Erp, 2011). The Dutch policy also meant that each and every sanction imposed in the market was publicized, instead of choosing prominent cases to focus on that might truly engender public outrage (Van Erp, 2011).

Of course, shaming may still prove ineffective in cases where people prefer to avoid information that makes them feel negatively (Loewenstein et al., 2014; Sunstein, 2020). Yet without a moral dimension to the publication, which would condemn certain behaviors of specific (and not all) entities, inclusive modes of disclosure like the Dutch policy would largely function as an ineffective "naming without shaming" (Van Erp, 2011). Van Erp further contended that these inclusive modes of enforcement publications actually do more harm than good, weakening the regulator in the eyes of regulatees while also creating a sense of governmental arbitrariness.

An additional point of view was offered by Fung et al. (2007), who argued that forms of "targeted transparency" are often incomplete, incomprehensible, inaccessible, and irrelevant to stakeholders. The authors use the targeted transparency framework to refer to mandated public disclosure of standardized and comparable information regarding specific products or practices to further a defined public purpose (Fung et al., 2007: 6). This framework encompasses regulatory disclosure schemes that may be regarded as RS, alongside other disclosure schemes relating, for example, to national security threats, campaign contributions, and criminal law in the noncorporate context, such as disclosure of sex offenders' place of residence. These policies are mainly meant to inform individual decision-making and warn people about various business, social, and political risks and to induce people and companies to improve compliance, performance, and behavior. While targeted transparency and RS are conceptually different, they do have a degree of overlap, for example, in corporate rankings.

Fung et al. (2007: 178) note that targeted transparency schemes do not always appropriately balance accuracy and comparability. That is, they do not always allow users to easily compare products or services, because they are overloaded with too much information; or, conversely, they tend to oversimplify information, thereby losing important nuances and even becoming misleading. For example, in the case of a mandated star-rating disclosure system relating to the rollover risks of sport utility vehicles (SUVs), the information was presented clearly in the form of a simple scale that facilitated comparison. Yet this scale gave a false positive impression of safety, because one star still represented a 40 percent chance of rolling over, which is rather high (Fung et al., 2007: 196).

Some of the problems, drawbacks, and risks associated with RS can be lessened by applying administrative safeguards, which some regulators already implement. In this context, several safeguarding steps have been proposed for policymakers: utilizing regulatory impact assessments to evaluate shaming costs versus benefits, including in comparison to other regulatory tools; conducting hearings before publications; taking privacy measures that allow targeted companies to restore their good name after a certain period of time (for instance, by deleting the information); consulting with the public and the regulated sector before introducing new shaming schemes; developing transparency mechanisms, such as guidelines that improve legal and regulatory certainty for businesses; and warning companies prior to shaming, giving them the opportunity to improve (Yadin, forthcoming-a).

Other works focus on elements relating to the policies themselves, and stress the importance of designing efficient and effective disclosure and shaming schemes that respond well to the problems discussed (Fung et al., 2007; Fung & O'Rourke, 2000; Sunstein, 2020; Van Erp, 2011; Yadin, 2019c). For example, Van Erp (2011) has suggested that regulators focus on demonstrating and exposing the harmfulness of carefully selected corporate behavior. Similarly, Fung & O'Rourke (2000) have advocated "populist maxi-min regulation," in which maximum public attention is given to minimal environmental performers, especially via blacklists and other rankings that can present technical data in a more shaming manner. These and other policy directions will be further discussed in Section 4.

2.2 Regulatory Climate Shaming

Section 1.1 discussed the significant deficiencies of extant climate regulation. Against this backdrop, this section will introduce an under-explored policy tool – the climate shaming of companies and industries by governmental regulatory bodies, which I will refer to as regulatory climate shaming (RCS). Regulatory climate shaming will serve as the central framework of this Element.

Drawing on the normative, policy, and theoretical dimensions of shaming, climate shaming, and RS, as well as relevant empirical evidence, this section will offer a conceptual and normative analysis of RCS, evaluating its unique rationales, advantages, justifications, and challenges. The main argument will be that shaming companies is a suitable, feasible, and necessary regulatory tool for fighting climate change.

Regulatory climate shaming refers to conveying information and/or messages to the public by national and subnational regulators in the executive branch, regarding corporate actions, omissions, decisions, and characteristics that have a harmful impact on climate change, in order to slow climate change and deter companies from continuing business as usual. The main goal of RCS is to induce companies to comply with climate-change norms (and to go "beyond-compliance"), using companies' sensitivity to their reputation among various stakeholders and to social pressure.

The idea is that companies that wish to avoid being named or presented as contributing to climate change (or even as being insufficiently climate-friendly) will adjust their actions so as to refrain from reputational damage that may translate into financial damages in the short or long run, in a temporary or a permanent manner. Commendatory publications may also play a role in regulatory climate-shaming schemes, using "naming-and-faming" tactics to recognize and promote climate-friendly business practices.

Regulatory climate shaming is also premised on the growing public awareness and concern regarding climate change and its dramatic implications, as people gain greater appreciation of the importance of the issue and of the risks it poses to basic human needs and values, including to the safety of entire cities and populations and of future generations (discussed in Sections 1.1 and 1.2). Regulatory climate shaming is also founded on people's strong views and emotional responses to the climate crisis (Leiserowitz et al., 2019), and on the ability and need to channel these into effective climate action via private enforcement mechanisms, especially given the failures of climate regulation in international and national arenas. The potential feasibility of RCS is reflected, for example, in a recent report by Yale University, suggesting that about a third of surveyed Americans have either rewarded or punished companies by buying or refraining from buying products, in response to those companies' climate-change actions and policies (Leiserowitz et al., 2021b).

Regulatory climate shaming is further based on the informational gaps that characterize the problem of corporate contributions to climate change and on the potential ability of regulators to capture the public's attention with new and important information in this regard. According to Leiserowitz et al. (2021b), most respondents say that companies should be doing more about climate change and that they would like to punish more companies for opposing climate action, but that they do not know which companies to punish. Surveyed participants also said that they would punish more companies, but that nobody has ever asked them to. In the context of consumers, as well as other stakeholders, RCS aims to meet this need for actionable information and utilize it to fight climate change.

In principle, RCS can be carried out via various schemes: for example, by creating a public governmental database with information on companies' greenhouse gas emissions, and indications of increases or reductions in these emissions over time; by designing and implementing a carbon-rating and labeling system for products, infrastructures, and services; by publicizing regulatory rankings and league tables of oil and gas companies and companies in other sectors according to indicators of their contribution to climate change; by publicizing the details of climate litigation cases and enforcement actions brought against companies; by lauding firms that are voluntarily reducing greenhouse gas emissions and adopting climate-friendly practices; and by posting condemnatory messages on social media and on governmental websites, naming and shaming firms for their climate-related behavior. More examples are provided in the next sections.

Within the RCS framework, stakeholders are meant to pressure firms by undermining their social license, thereby improving compliance and strengthening climate social norms relating to corporate behavior. The mere threat of public pressure is also meant to function as deterrent of undesired climate actions and decisions of firms.

Private sanctioning can also be carried out indirectly by stakeholders in response to RCS publications, by preferring to support and engage with firms that are lauded (in those publications or in subsequent coverages) as contributing to climate mitigation and adaptation. Indications of the feasibility of regulatory climate faming can be found in recent American and European surveys, in which most respondents not only said that they intend in the near future to punish companies that oppose climate action by not purchasing from them but also that they will purchase more goods and services from companies that adopt climate-friendly policies (Leiserowitz et al., 2021b; Leiserowitz et al., 2021c).

Regulatory climate shaming can shame specific companies or facilities, groups of companies or facilities, or even entire sectors. Importantly, RCS is not restricted to targeting fossil-fuel companies, as other industries – such as transportation, construction, manufacturing, energy (electricity), retail, finance, agriculture, infrastructure, and advertising – are also contributing to climate change. Many industry sectors are currently highly dependent on fossil fuels and use them extensively to operate facilities and transport commodities (Fischedick et al., 2014). Companies in diverse sectors also contribute to the total emission of greenhouse gases by using dirty energy sources (such as coal, oil, and gas) to support their supply chains – for example, through the purchase of goods and services, through waste disposal, and even through the use of their products by consumers (for instance, people driving cars) (Hertwich & Wood, 2018).

Many companies and financial institutions also invest in the fossil-fuel industry and in other industries that have yet to adopt climate-friendly policies, or may advertise such industries, thus contributing indirectly to greenhouse gas emissions and exacerbating the climate crisis. Again, these issues already strike something of a chord with public perceptions: one in three Americans say that they would switch banks if they knew that their bank was investing in fossil-fuel companies (Leiserowitz et al., 2021b). **Figure 3** details these and other possible stakeholders' responses to RCS.

Admittedly, "shaming" is a rather provocative term, yet in this context it properly captures the deep disapproval that corporations' contributions to climate change deserve, including their use of manipulative tactics to impede climate action, which will also be discussed in this section. I choose to use the term "shaming" not only because it implies that corporations are responsible for climate change but also to underscore that they are considered by governments and regulators to be key players in climate-change mitigation, and that they are further expected to react to these reputational sanctions with measures that will decrease greenhouse gas emissions. Other terms, such as "transparency," "disclosure," or even "reputational sanctions," do not carry the same weight of meaning, assignment of responsibility (or even blame), expression of disapproval, and nudge to action (of both corporations and the public) as does "shaming."

Indeed, it is possible that regulators who structure and implement information-sharing schemes relating to climate change have not intended or are even not fully aware of the inherent shaming effect of these publications, nor of the possible behavioral change that such disclosure action may bring. The regulatory climate-shaming framework may prove especially useful in these situations, bringing to center stage the shaming mechanisms, processes, and functions that are in effect present in such policies.

As will be discussed in greater detail in the next section, RCS is at a relatively preliminary stage in policy domains. It currently operates via mechanisms such as naming and shaming firms for noncompliance with climate-change laws and

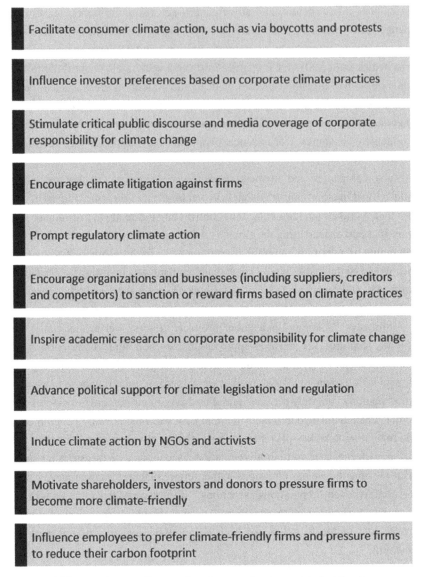

Figure 3 The regulatory climate-shaming mechanism: stakeholders' responses

regulations; ranking companies in blacklists that are based, among other factors, on their contribution to the climate crisis; requiring climate labels on products and buildings; mandating financial and consumer disclosure about climate risks; producing online databases that allow users to view companies' emission data; and "climate faming" of companies that join voluntary programs or adopt voluntary standards.

In addition to the general advantages of RS discussed in Section 2.1, several unique advantages can be attributed to RCS. As discussed in Section 1.1, climate law, regulation, and governance at multiple levels and across multiple jurisdictions are generally proving unsuccessful in mitigating climate change and reducing emissions at a sufficient rate. Against this backdrop, RCS can be especially useful in meeting the increasing need to develop innovative and effective climate policy tools. While regulators are grappling with assembling their regulatory toolkits, scientific evidence of the urgency and gravity of the situation is accumulating, and many people are already suffering the results of global warming. Climate shaming could therefore be inserted into the climate policy domains at a crucial time.

In line with various regulatory theories and research, including RS studies in fields closely related to climate change (see Section 2.1), RS has the potential to enhance or replace other climate regulation tools and strategies. According to responsive regulation theory (Ayres & Braithwaite, 1992), a combination of a variety of soft and hard regulatory tools is needed for improving corporate compliance and increasing deterrence. In the field of environmental policy, smart regulation theory (Gunningham et al., 1998) similarly suggests that utilizing multiple rather than single policy instruments as well as a broad range of actors, including third parties, can yield better regulatory results. Studies of RS have further shown that shaming companies for poor compliance can function as a substitute for command-and-control in terms of general and specific deterrence in the field of public health (Christensen et al., 2017; Johnson, 2020).

Similarly, the combination of innovative, soft, public-based tools like shaming with existing, more paradigmatic regulatory tactics like command-and-control can potentially create an integrated strategy for effectively addressing the climate crisis (Dessler & Parson, 2019; IPCC, 2022b). Climate shaming could therefore offer an important and necessary soft-law private enforcement tool in the climate enforcement pyramid, to be utilized in combination with other hard and soft climate regulation tools. Climate shaming may also play an important role in jurisdictions in which some hard-law climate tools, such as strict limitations on emissions, are unavailable to regulators for various reasons (see Sections 1.1 and 2.1), including legislative deadlocks and unfavorable court rulings. For example, the US Supreme Court recently ruled that the EPA cannot issue broad climate-change rules to regulate power plants.[16] Regulatory climate shaming could serve as an experimental tool as well, prior to enacting mandatory climate obligations, which are generally harder to legislate.

---

[16] See *West Virginia v. EPA* (2022).

It should also be noted that other regulatory measures in the realm of corporate enforcement do not sufficiently express society's condemnation of corporate conduct that exacerbates the climate crisis, as shaming does. Many companies worldwide have emitted and continue to emit greenhouse gases, despite indisputable scientific evidence of their effect on climate. Indeed, Frumhoff et al. (2015) emphasize that more than half of all industrial carbon emissions in history have occurred since the formation of the IPCC and the establishment of climate-change science. Importantly, business alternatives are available to firms, such as altering business models toward renewable energy and investing in low-carbon technologies, but many firms nevertheless choose to continue with a "business-as-usual" approach (Frumhoff et al., 2015). It can therefore be argued (in line with the terminology discussion above) that shaming better reflects society's disapproval of corporate contribution or indifference to climate change than monetary sanctioning.

Certainly, as explained in the previous section, hard law does not always provide sufficient incentives for corporations to alter their behavior. Even criminal law measures usually result in companies, which naturally cannot be incarcerated, simply paying a price tag for their violations in the form of fines. This concern is especially true for major corporations like the fossil-fuel companies ("carbon majors") and other big companies with a substantial carbon footprint, which are hard to deter via direct monetary sanctions. Regulatory climate shaming can overcome these limitations of hard law and promote climate-responsible behavior by corporations based on other motivations. For instance, companies may want to improve their standing in regulatory publications reflecting their level of compliance with climate norms – such as blacklists, databases, rankings, labels, and reports – driven by fears of backlash from various stakeholders. This is a much wider system of deterrence, because it is based on a variety of third parties (and not only on those who are in direct contact with the firm), and on more general information such as total emissions rather than solely on regulatory violations. Examples of such climate-shaming schemes are provided in more detail in Sections 3 and 4.

In this vein, recent studies have indicated that various policies involving mandatory public disclosure of firms' greenhouse gas emissions lead to a decrease in firms' emissions in different jurisdictions. These studies suggest that such schemes have a "pillory effect," based on firms' social license among various stakeholders, including shareholders. For instance, Downar et al. (2021) show that firms have reduced emissions by 8 percent following a 2013 UK requirement for publicly listed firms to disclose their total greenhouse gas emissions in their annual financial reports. Similarly, Tomar (2022) shows that US facilities also reduced their greenhouse gas emissions by some

8 percent following the introduction of the EPA mandatory greenhouse gas disclosure program and the publication of the data in 2012 via open online databases. Other studies also point to improvements in carbon intensity following the introduction of these greenhouse gas mandatory disclosure rules in the United States and the United Kingdom (Bauckloh et al., 2023; Jouvenot & Krueger, 2019).

Other climate disclosure tools with a shaming effect, such as disclosure of the use of fossil fuels versus renewables in electricity bills sent to consumers, have also improved firms' climate performance (Delmas et al., 2010). In addition, various studies point to the potential of carbon labels to influence various corporate stakeholders, such as consumers, suppliers, retailers, transporters, and producers, and drive corporations to adopt climate-friendly practices (Taufique et al., 2022).

Companies may also wish to avoid scoring or ranking low in regulatory lists, rankings, databases, and other types of publications that are based on companies' participation in voluntary climate programs. Generally, voluntary climate programs are run by both industries and governments – the EPA alone operates dozens of voluntary climate programs at federal and state levels (Freeman, 2021; Hsueh & Prakash, 2012) – and firms may be induced to join such programs for a variety of reasons (Berliner & Prakash, 2013; Hsueh, 2020; Prakash, 2000). For example, the EPA operates the Methane Challenge program,[17] which aims to encourage oil and gas companies to reduce emissions of the greenhouse gas methane. In exchange, reputational gains are offered to participating companies, through prominent publication in newspapers and industry journals. Combining voluntary climate programs with a regulatory shaming and faming approach may serve as an additional nudge for firms to join such programs: companies that do not participate in any voluntary climate program, for example, may suffer shaming, especially when compared to other companies that are more involved with voluntary climate programs, via the use of ranking, grading, lists, and the like.

Participation in voluntary programs can also be a valuable strategy for offsetting reputational injuries for companies that have been previously shamed by regulators, either for legal noncompliance or for poor climate practices. Research suggests that CSR activity offers corporations a type of "insurance" against reputational harm generally associated with regulatory and legal actions taken against them (Godfrey et al., 2009). Similarly, RCS may nudge companies to join climate voluntary programs to counteract previous or future reputational damage caused by RCS.

---

[17] See www.epa.gov/natural-gas-star-program/methane-challenge-program.

Regulatory climate shaming based on publication of violations could also prove effective in inducing compliance. Indeed, the conventional wisdom holds that generally regulatory publication of environmental infringements does not cause reputational harms in capital markets, meaning that any drop in share prices tends to reflect no more than the financial penalties incurred by the firm (Brady et al., 2019; Karpoff, 2012; Karpoff et al., 2005; McGuire et al., 2022). However, these studies mostly relate to environmental rather than climate issues, and mainly analyze incidents and market behavior from over a decade ago, when public sensitivities to environmental issues were weaker.

The potential of shaming to induce firms to react quickly and change their business practices may also prove especially suitable to tackling climate change – a problem that companies can arguably easily ignore and neglect for many more years. While net-zero commitments by states and firms, which generally relate to substantial reductions in greenhouse gas emission over a period of five to thirty years, have been described as a "burn now, pay later" approach, RCS may be considered a "burn now, pay now" strategy. That is, it can inflict reputational costs on companies for continuing business as usual, within a time frame that is more suitable to the situation. The relatively low costs of RCS can also contribute to its rapid implementation, thus providing a speedy response to an urgent problem.

Importantly, RCS can also serve as an especially suitable reaction to the indirect contributions of firms to the climate crisis via climate obstruction. This is not to imply that RCS should be used as a retaliation tool or as punishment (see Section 1.3). While a full moral argument could be made against the fossil-fuel industry and other industries, the argument advanced here is more practical in nature, viewing RS as an enforcement tool that can slow climate change.

Currently, there are large numbers of firms that not only produce emissions, and/or finance, depend on, or advertise the fossil-fuel industry, but also intensively work to thwart climate regulation and climate action more generally, including by using manipulative practices such as climate denial and climate washing (Yadin, forthcoming-b). These climate-obstruction practices seek to set back climate-change regulation efforts and legitimize a business-as-usual approach, thereby contributing to the continuance of greenhouse gas emissions and exacerbating the climate crisis through sophisticated indirect methods.

Generally, climate denial is the rejection of climate-change science based on opinion, ideology, emotions, or interests (Washington & Cook, 2011). For many years, large oil and gas companies have been funding research and running public campaigns to establish and promote climate denial, for their own financial interests (Michaels, 2020). Beginning in the late 1980s and throughout the 1990s and 2000s, the oil and gas industry utilized various front groups and think tanks to formulate climate-denial strategies that would combat any regulatory endeavors posing a threat to the industry's bottom line (Freese, 2020). Notably, leading fossil-fuel corporations formed organizations whose goal was to oppose policies designed to reduce greenhouse gas emissions and litigate against environmental regulators, challenging their authority to regulate greenhouse gases (Frumhoff et al., 2015). Internal documents revealed in recent years prove that these strategies were being pursued even as companies' own internal research pointed very early on to the catastrophic climatic effects of greenhouse gas emissions (Freese, 2020; Supran & Oreskes, 2021).

The oil and gas industry has also employed lobbyists, media consultants, and public relations companies to establish its narrative of denial, spread misinformation, and influence public opinion (Almiron & Xifra, 2020). Common denial messages focus on denying the scientific evidence of climate change; minimizing climate-change effects and implications; denying the anthropogenic (human-caused) origins of climate change, and instead pointing to nature itself as the cause; arguing that nothing can be done to mitigate climate change; and arguing that science and technology will solve the problem eventually, so nothing should be done now (Almiron & Xifra, 2020; Freese, 2020).

Climate-change denial by the oil and gas industry, often described as a "denial machine," is considered to have played a major role in setting back climate-change mitigation efforts (Freese, 2020). It is often compared to the efforts of the tobacco industry to cover up and actively deny the devastating impacts of smoking on public health despite evidence to the contrary, including from those companies' own data, internal reports, and scientific studies (Michaels, 2008, 2020; Supran & Oreskes, 2021). It is against this background that scholars have suggested that climate-denial tactics employed by the fossil-fuel industry should be considered criminal (Haines & Parker, 2017; Kramer, 2020).

Today, while many oil and gas companies have altered their climate-denial tactics, moving from attacking the science to employing delaying tactics that focus on denying the urgency of the situation, and to blame-shifting tactics (e.g. by arguing that consumers are to blame, for depending so deeply on fossil-fuel energy), these efforts continue to impede climate action. Meanwhile, litigation brought against the industry based on climate-denial practices (see Section 4) is lagging and uncertain, sometimes culminating in the fossil-fuel companies winning in court, arguably in light of the generally underdeveloped nature of climate law (Freese, 2020; Yadin, forthcoming-b).

In many respects, RS may function as a uniquely equipped tool to advance climate regulation goals by fighting manipulative climate-obstruction tactics (Yadin, forthcoming-b). This is because it harnesses credible information sharing by the government to combat corporate disinformation and deception, while publicly assigning blame and liability to industries and companies that often deny such blame or seek to shift it elsewhere (to consumers, for example). Also, RS has the potential to become a highly suitable means for addressing climate denial because both shaming and denial are communication-based strategies that address corporate reputations and aim to influence public opinion. While climate denial seeks to harness the public to pressure policymakers into maintaining fossil-fuel dependence, RCS aims to harness the public to pressure corporations and policymakers into decreasing our reliance on fossil fuels.

Regulatory climate shaming can also be highly relevant for addressing the climate-washing tactics employed by firms. Companies now flaunt unverified and misleading private climate ratings, assigned by unregulated ESG (environmental, social, and governance) rating agencies, and publish their own misleading climate statements, labels, and reports in order to gain public recognition and improve their reputations (Hsueh, 2020; Li et al., 2022). Importantly, climate-washing practices are not only misleading but they also inhibit climate-change mitigation by allowing firms to evade the reputational, social, and environmental costs of their actions. Regulatory climate shaming can disseminate reliable information on corporate climate performance, and specifically on corporate climate-washing practices, rendering it a much less appealing business strategy. This type of RCS can also protect the personal autonomy of consumers, investors, and other stakeholders, giving them the ability to act in accordance with their climate preferences, morals, politics, and world views, without being manipulated by corporations in the informational arena.

While not a primary goal, RCS can also help educate the public and the industry about climate change. It can inform the public on specific aspects of climate change (Brooks & Ebi, 2021), such as the sources of greenhouse gas emissions, the effect of those emissions, and the implications of climate change. More generally, RS mechanisms can send a signal to the general public about the importance of climate change, as well as to various stakeholders, including business actors, and strengthen climate change norms. The educational benefits of RCS schemes may even transcend their functionality in the shaming mechanism and encourage people to become more climate-active in various other different ways.

Regulatory climate-shaming schemes can also influence consumer preferences and encourage behavioral changes in individuals to create an aggregated effect of greenhouse gas emission reductions at the end-user level. Thus, information given at gas stations, on electricity bills, and on airline tickets can not only shame companies into direct emission reductions (e.g. by implementing new technologies) but also nudge consumers toward climate-friendly choices, for example, by scaling back traveling and heating. A related effect of RCS schemes involves supporting decision-making (without nudging) – for example, by providing more relevant information to customers considering the purchase of a large vehicle, in addition to information on costs, performance, and suitability for specific needs.

Information revealed via RCS schemes also promotes governmental transparency relating to climate action, allowing the public to monitor and criticize municipal and national policies and activities, pressure policymakers to take proactive climate action, and become informed voters. It can also indirectly promote the public right to know regarding corporations that may be responsible for risks such as floods or fires, and for municipality and state expenditure relating to adaptation measures (such as investment in necessary infrastructures). Regulatory climate-shaming schemes that are based on the publication of enforcement actions such as civil penalties can also promote regulatory transparency about sanctioning of firms, and advance legal and regulatory certainty regarding climate enforcement policies.

Finally, RCS can also help regulatory bodies improve their own public image and gain greater trust and respect from various stakeholders. Against the failure of classic regulatory tools to combat climate change (see Section 1.1), RS may help regulators regain their status in the eyes of the public and appear active, relevant, and resourceful.

Shaming could therefore prove to be a beneficial, efficient, and effective regulatory tool in the fight against climate change. Its strong rationales, justifications, and theoretical background offer regulators an important policy avenue that should be further developed. Certainly, the use of shaming tactics by regulators may be regarded a highly controversial practice, incorporating as it does public condemnation of private business organizations. However, the severity of the climate crisis and its extreme, wide-ranging implications for public interests and human rights, combined with the current complex limitations of more conventional climate regulation (see Section 1.1), warrant the adoption of somewhat unconventional measures such as shaming.

Both parts of this section have shown that shaming could equip regulators with new capabilities on the climate-change front, by harnessing public opinion and public action. A deeper look into the practice of naming-and-shaming policies implemented in related contexts and discussed in both parts of this section revealed that RCS is in fact not as unconventional as one might think, as it is currently being successfully used by many public health and environmental protection policies worldwide. However, the RS approach may also pose several specific challenges in the climate-change context that should be acknowledged and addressed (in addition to the more general challenges posed by RS discussed in the previous section).

For instance, since climate change is a highly controversial and political issue (Leiserowitz et al., 2022) – despite the fact that the science is very clear – RS may backfire. Namely, regulators who engage in shaming may appear to be politically rather than professionally motivated and lose public credibility. Some people (and organizations) may even want to defy RS and support shamed entities, for example, by purchasing their product or investing, and react aggressively toward the shamers (regulators and public members). Some jurisdictions may also create counter-lists, shaming companies that intend to withdraw from fossil fuels (Hagan & Brush, 2022).

This effect is dependent, of course, on cultural, economic, environmental, and political perceptions, which may vary among jurisdictions. However, regulators will have to face, sooner rather than later, the implications of implementing a variety of climate policies that may also result in backlash from various stakeholders. Furthermore, critiques regarding the potential indirect costs of shaming – such as the loss of jobs in the fossil-fuel industry, shortages of energy sources, and energy price increases – are also not unique to shaming and apply equally to various other climate regulation tools that aim to curb greenhouse gas emission (Choudhury, 2021).

Despite these caveats, climate shaming as a policy tool has a good chance of gaining public legitimacy and support because of the current context, in which extreme climate events that cause irreparable harms are increasing in scope and in frequency and at the same time climate law and regulation are recognized as extremely insufficient (see Section 1.1). Fung et al. (2007) suggest that information-based policies can overcome various political hurdles – including policy obstruction by industry members and other stakeholders – in times of crisis (e.g. financial or health crises), when the public is highly aware of the public interest being harmed. They argue that in cases of disasters or catastrophes, policymakers are able to more easily gain public legitimacy for enacting broad disclosure mandates, as the public recognizes that current laws and regulations do not appropriately prevent risks and that new tools must therefore be utilized. Thus, it can be argued that the strong public sentiment regarding the climate crisis could help overcome various impediments to implementing RCS.

It is true that RCS may result in costs to the government if it becomes the subject of litigation, rendering such policy less efficient and effective, especially considering the vast resources of the oil and gas industries and their use of climate-obstruction tactics. However, it should be much more difficult for companies to attack RCS for being politically biased, unfair, overly burdensome, or uninformed when it is mostly carried out by the companies' own consumers, investors, customers, and the like. Additionally, such litigation may harm the reputation of litigating companies, including in the fossil-fuel industry, which are currently making an effort to (at least) appear to be climate-conscious, if not climate-friendly (Li et al., 2022). That is, such litigation may be effective in thwarting climate-shaming regulation, but not necessarily in pleasing stakeholders such as investors, financers, and consumers, nor in satisfying policymakers. A shaming approach may also be generally regarded as fairer by firms, and thereby reduce corporate climate litigation brought against regulators. This is because RCS does not forcefully dictate new emission standards, require the implementation of new technologies, or ban certain products or business activities, but instead uses nudges, information sharing, and private enforcement tools to mitigate climate change.

On the other hand, it could also be argued that crowdsourced enforcement such as shaming is not an appropriate approach in cases of extreme and immediate danger, as presented by the climate crisis, which call for fierce governmental intervention. In a similar vein, the costs and responsibilities entailed in climate enforcement by shaming could be viewed as unfairly shifted from the government to the public, who function as surrogate regulator, and from corporations to private individuals. One could also claim that the use of shaming may induce governments to stop trying to legislate necessary hard-law command-and-control climate regulation, and unwisely gamble on the success of soft-law private governance mechanisms.

However, it is important to remember that RCS does keep regulators involved in initiating, supervising, and managing the process, and so this approach should be understood as a public–private partnership, rather than a wholly privatized activity. This public–private enforcement mode can also help overcome the problem of insufficient regulatory budgets and personnel, which is prevalent in many regulatory agencies worldwide, especially in environmental and climate agencies and departments (Konisky & Woods, 2016, 2018). Yet it should be emphasized that RCS divides the costs and efforts of entrenching and enforcing climate norms among a large number of people who are willing and able to participate in the shaming process, without forcing anyone to take part. In this way, the regulatory task is split into very small portions, which do not overburden shaming agents but together can add up to substantial regulation in quantity and quality.

Furthermore, the idea that command-and-control alone can effectively address the challenges in today's markets, and specifically global environmental issues, has long since been abandoned. Soft regulatory tools such as information-based, voluntary, and consensus-based policies have in many cases proved to be more effective (Ayres & Braithwaite, 1992; Baldwin et al., 2012; Gunningham & Sinclair, 2017). In light of the failure of existing approaches to climate law, regulation, and governance discussed in Section 1.1, it seems appropriate and justified to explore new policy avenues such as climate shaming.

Some may also argue that certain industries, and specifically the fossil-fuel industry, are beyond shaming, or shameless. However, the existence of the "climate-denial machine" and of climate-washing discussed in this section indicates that corporations, especially fossil-fuel companies, are sensitive to their public image. Whether or not this sensitivity is based on nothing more than concern for the bottom line is less important than the fact that it exists.

Others might argue that specific companies cannot be shamed if entire sectors are acting with the same lack of regard to climate change. Yet, as companies in, for example, the automobile industry, the financial sector, or the advertising industry compete with one another for clients and investors, it is possible to shame a specific company even when other companies in the same marketplace share the same practices and avoid adopting climate-friendly policies. Additionally, entire sectors can be shamed en masse (examples are discussed in Section 3.1), jeopardizing their current (convenient) regulatory landscape. In these situations, industry leaders may be inclined to react to regulatory climate-shaming publications and improve business practices, thereby encouraging other firms to follow the same path.

Another challenge, generally discussed in Section 2.1, relates to companies trying to manipulate regulatory climate-shaming schemes, for example, by making superficial changes in their operations that would improve their ratings, presented data, scores, or grades but without instituting meaningful overall greenhouse gas reductions. Indeed, a recent study indicates that firms that own multiple plants reduce greenhouse gas emissions in plants subject to mandated disclosure of emissions, while increasing emissions in plants that are not covered by such disclosure rules (Yang et al., 2021). In this vein, companies may also join climate programs just to be named and famed, without fully complying with the terms of the schemes (Prakash & Potoski, 2006). Such practices may be developed in particular by industries such as oil and gas, as an extension of their current climate-obstruction practices.

To support the development of effective, well-balanced, sustainable, and legitimate regulatory climate-shaming schemes that minimize the risks and challenges discussed in this section and fully utilize shaming's potential to mitigate climate change, the next sections will focus on policy design. To that end, Section 3 will first provide an analysis of extant regulatory climate-shaming practices in various forms and jurisdictions, and then Section 4 will suggest ways to adapt and develop this regulatory approach.

3 Regulatory Climate-Shaming Schemes

This section presents prominent climate-shaming schemes currently being developed and executed in several jurisdictions. The aim of the section is to illustrate the different ways in which RCS currently operates and to spotlight its extant features and mechanisms, in order to form a descriptive theory that will shed light on appropriate directions for policy design. More specifically, this section provides a contemporary snapshot and a typology of regulatory climate-shaming schemes by surveying examples from four different jurisdictions which demonstrate prominent use of the approach: European Union member states, the United States, the United Kingdom, and Israel. Examples are given of mechanisms anchored in laws, directives, rules, regulations, local laws, decrees, notices, memorandums, and the like (see **Appendix A** online[18] for further details), at both national and subnational policy levels.

Regulatory climate shaming is a growing practice, with many different types of schemes beginning to emerge. However, in many of the schemes presently operating in surveyed jurisdictions, shaming functions as a secondary goal to the primary goal of the act of publication, which may be educating, supporting decision-making (informing), warning, protecting the public's right to know, promoting legal certainty, nudging individuals, or promoting governmental transparency (**Figure 1**). Other, more explicit forms of RCS underscore the negative aspects of corporate climate performance, using, for example, blacklisting, ranking, scoring, rating, and publication of regulatory violations and enforcement actions. Generally, in these types of information-sharing schemes the shaming of firms as a regulatory tool functions more prominently.

Additionally, the climate portion within some shaming schemes is obscure, embedded in regulatory goals relating to fields such as environmental protection and CSR. Conversely, other climate-shaming schemes are highly focused on promoting mitigation of climate change. Several new shaming policies that are currently being considered by surveyed regulators but have not yet been formally introduced also include a clear and central climate element.

In addition to the expansion and development of regulatory climate tools based on negative publications ("shaming"), positive publications ("faming") of firms' climate-related practices are also proliferating. Combined positive and negative publications on corporate performance relating to climate change are also evident, for example, via rankings and lists. Notably, surveyed policies tend to be based on voluntary norms. That is, firms are mostly shamed for underperforming in the ethical, moral, social, and environmental sense, rather than for legal violations. Climate faming focuses on these domains as well.

---

[18] www.cambridge.org/yadin_resources.

The study of these different policies further reveals that, like most RS policies, RCS practices are usually not officially presented as shaming. Instead, they are typically framed by policymakers as transparency policies (see **Appendix A**).[19] In practice, other stakeholders, institutions, and organizations – such as NGOs, industry actors, researchers, lawyers, scholars, and the media – often highlight the shaming elements of these policies, either critically or positively.

This section also reveals that information disclosed by surveyed regulators on corporate climate performance is sometimes structured in a way that shames an entire industry sector, rather than a particular firm. Shaming tactics also differ in the sense that some involve publication activities carried out by the companies themselves (e.g. labeling schemes), while others comprise publication activities by the regulator (e.g. blacklists). Some recently introduced or updated climate-shaming schemes are based on new digital capabilities. These sometimes combine publications by both firms and regulators, such as QR codes stamped on product labels which refer consumers to regulatory databases.

3.1 Prominent Shaming Schemes in Surveyed Jurisdictions

This section explores prominent climate-shaming schemes in European Union member states, the United States, the United Kingdom, and Israel, and presents them according to several different categories. This categorization is not absolute, and some overlap between categories is possible. **Table 1** and **Appendix A**[20] provide more detail and references.

Publication of Enforcement Actions

Within this mode of shaming, regulators publicize the names of companies that have breached their climate obligations and were legally sanctioned. For instance, the UK Environment Agency posts information regarding civil penalties imposed under all UK climate-change laws and regulations on a governmental website.[21] The Agency's publication comprises several lists that specify the company name next to the sum of penalty imposed and provide a description of the infringement. According to the Agency, these penalties are published for a minimum of twelve months. The lists are rather short, which can be more shaming than long lists, as company names stand out more. In fact, one such list published in 2022 comprises only one company, detailing seven different infringements (thereby increasing the shaming effect). However, the information is generally very technical in nature, and is not presented in an easily accessible or communicable form.

---

[19] www.cambridge.org/yadin_resources.
[20] www.cambridge.org/yadin_resources.
[21] See note 2.

Additionally, the Agency publishes lists with company-specific information on performance and breaches of voluntary climate agreements made between the Agency and specific firms, including civil penalties imposed. Typically, such climate agreements in the United Kingdom involve companies agreeing to reduce carbon emissions in exchange for carbon-tax deductions. The Agency publicizes information on breaches of these climate agreements, as well as data relating to targets set in the agreements and whether they were met and total company emissions. In addition to the potential for shaming specific firms, emission-reduction commitments in sectors such as food, textiles, wood, steel, and plastics are also published by the Agency, creating a sectoral shaming effect for poorly performing sectors. However, these datasets are also not in an accessible form (presented in detailed tables), and are devoid of context, explanations, or clear messaging.

Similarly, many European Union member states (and the United Kingdom) publish the names of companies that have breached their cap-and-trade obligations under the terms of the European Union Emissions Trading Scheme (EU ETS) (Fleurke & Verschuuren, 2016; Luna, 2019). Generally, the EU ETS covers some 10,000 installations, which account for around 40 percent of the EU's greenhouse gas emissions (European Commission, n.d.-a). The ETS directive instructs member states to not only fine but also publish the names of operators and installations in the power sector, manufacturing industry, and airline industry which have breached the duty to surrender sufficient allowances to cover their emissions. However, not all member states have implemented this publication mechanism, and those that have publish the names of noncompliant companies in a way that has been criticized in scholarly works as inaccessible (using short textual data inside reports, posting spreadsheets on agencies' websites, or issuing publications in official gazettes) and therefore ineffective (Fleurke & Verschuuren, 2016; Luna, 2019).

Rating, Ranking, Scoring, and Blacklisting

Other forms of shaming are carried out in surveyed jurisdictions via rating, ranking, scoring, and blacklisting. A prominent example for this type of shaming is the Israeli Ministry of Environmental Protection's "red list of Israeli factories," which annually scores and ranks companies according to environmental performance, also taking into account their participation in voluntary private and governmental climate programs (Israeli Ministry of Environmental Protection, n.d.). According to the Ministry's scoring methodology, companies' voluntary compliance via climate programs can offset their adverse environmental data and improve their total score. Rankings are posted on the Ministry's website and social media accounts,[22] and circulated as press releases. Each company's detailed score and data, including participation and nonparticipation in voluntary climate programs, are also published on the Ministry's website.

Another example involves the ranking of car models offered for sale or lease in EU member states. In accordance with Directive 1999/94/EC, a list of all new car models offered for purchase or lease and their CO_2 emissions (and fuel consumption) must be prominently displayed at points of sale. According to the Directive, car models are to be grouped and listed separately according to fuel type (e.g. petrol or diesel etc.), and within each fuel type, models should be ranked in order of increasing CO_2 emissions, with the model with the lowest fuel consumption being placed at the top of the list. The text "CO_2 is the main greenhouse gas responsible for global warming" must also be displayed next to the list.

Shaming schemes via rating, ranking, scoring, and blacklisting are also currently being developed by banking regulators. For instance, due to significant noncompliance with climate-risk disclosure rules, the European Central Bank (ECB) is considering publicly listing banks that engage in climate washing or that repeatedly fail to fully disclose their climate risks (Arnold, 2022). Another case in point is that of the Bank of Israel's Banking Supervision Department, which plans to publicly grade and rank Israeli banks according to climate-risk indicators (Ashkenazi, 2021).

Climate Labels

Another method of climate shaming by regulators involves the use of climate labels. Generally, climate labels can not only educate, inform, and nudge consumers (Brooks & Ebi, 2021; Cohen & Viscusi, 2012) but also induce public pressure on firms and shame them into improving their business practices and limiting their products' carbon footprint (Taufique et al., 2022). Several climate labeling schemes focus on private transportation and fuel consumption. For instance, the Swedish Energy Agency requires companies to place labels on fuel pumps and points of charging in charging and gas stations, displaying company-specific climate-impact ratings for different fuels (Swedish Energy Agency, 2021). In accordance with Sweden's Fuel Ordinance, the label must contain information regarding the fossil raw materials or renewable raw materials included in the fuel, related greenhouse gas emissions, and the country of origin of the raw materials.

In another example, EU member states are required to label cars at the point of sale and in advertisements with details of their carbon emission (and fuel

---

[22] See, for example, Ministry of Environmental Protection (@SvivaMinistry), www.facebook .com/svivaministry.

efficiency), intended not only to inform and nudge consumers but also to encourage manufacturers to reduce the fuel consumption of new cars. The labels also include a statement explaining that CO_2 is the main greenhouse gas responsible for global warming.

Similarly, tires in EU member states are labeled with a color-coded score on a scale from A to E, which indicates fuel efficiency (among other qualities), as part of the EU's climate policy aiming to reduce greenhouse gas emissions. The labels also include the supplier's name at the top of the label and are intended, among other goals, to encourage manufacturers to innovate, improve their products, and gain a higher rating (European Commission, n.d.-b).

Similar labeling methods have also been taken up by countries and cities worldwide in the context of energy ratings for buildings (Taufique et al., 2022). The general idea of this type of labeling is to encourage contractors to build energy-efficient buildings and thereby reduce electricity consumption and promote energy independence. These labeling schemes are currently proliferating and becoming more visible, digital, and communicative. For instance, the city of New York has recently mandated building owners to post their energy efficiency rating label near each building entrance. The New York rating comprises a letter ranging from A to F, a numerical score ranging from 0 to 100, and a color-coded scale, in which D is colored red and A is colored green. Building owners are also required to publish, next to these indicators, the energy rating of the building for the previous two years, thereby highlighting improvement or regression. The city further posts the rating and scores of all relevant buildings on its website. For comparison, other surveyed jurisdictions merely require the publication of the rating to potential residents (Israeli Ministry of Energy, 2020). The EU Energy Performance of Building Directive, conversely, requires that energy performance certificates are also included in advertisements for buildings that are up for sale or rent.

Climate labels have also been enhanced by digital means such as QR codes and online databases, which allow consumers to search energy efficiency data on specific companies and products. Notably, the European Product Registry for Energy Labelling (EPREL) allows users to view energy labels for products such as white goods, electronic displays, and car tires, by scanning the QR code featured on the labels.

Other forms of climate labels consist of mandatory messages in companies' commercial campaigns and at points of sale, which can not only nudge and educate consumers but also shame industries with heavy-carbon-footprint products. For example, a new regulation from the French Ministry of the Ecological Transition requires automobile manufacturers to include in their advertising a message that encourages people to prefer public transport and cycling to driving, when possible, undermining the marketing message and the public image of the automobile industry more generally. The message can be one of the following three: "For short trips, prefer walking or cycling," "Think about carpooling," or "On a daily basis, take public transport." In a similar vein, the municipality of Cambridge, Massachusetts has passed a city ordinance that mandates the clear labeling of all fuel pumps, stating that burning gasoline, diesel, and ethanol has major consequences for human health and the environment, including contributing to climate change.

Climate Disclosure in Company Reporting

Other forms of climate shaming can be found in the climate-reporting obligations for publicly listed firms and financial institutions which are currently being implemented or developed by financial and energy regulators in all surveyed jurisdictions. Noticeably, the US SEC recently proposed a new rule that requires publicly traded companies to issue detailed climate-risk disclosures.[23] Under the proposed new rule, firms will need to disclose their level of reliance on fossil fuels, their direct and indirect greenhouse gas emissions, and whether they have a transition plan in place to deal with climate change. This type of climate disclosure may mandate car manufacturers, for example, to state in their filings that their reputation and stock price may be harmed due to greenhouse gas emissions from their vehicles. Generally, climate disclosure in company filings may shame companies into preparing climate transition plans, fearing public condemnation and other negative responses from, for example, investors, consumers, NGOs, and advertisers (Downar et al., 2021; Jouvenot & Krueger, 2019). Similar steps have also been taken by regulators outside of the financial disclosure landscape. For instance, French electronic communication operators, such as mobile phone operators and internet service providers, are required to publish their policies for reducing greenhouse gas emission.

Emission Databases

Regulatory climate shaming is also being pursued via emissions databases, which can not only inform stakeholders but also utilize companies' reputational sensitivities to induce emissions reductions (Bauckloh et al., 2023; Tomar, 2022). A prominent example is the US EPA's "Facility-Level Information on GreenHouse gases Tool" (FLIGHT), which enables users to view data in maps, tables, lists, charts, and graphs for individual facilities, groups of facilities, and sectors, and to compare emission trends over time.[24] FLIGHT generates, for example, rankings of firms based on their greenhouse gas emissions according to

---

[23] See The Enhancement and Standardization of Climate-Related Disclosures for Investors, 87 FR 21334 (2022).
[24] See note 1.

various search criteria, and provides graphs showing companies' increasing emissions over the years.

Climate Naming and Faming

Finally, RCS is also performed by various modes of climate faming, in which firms are lauded for their climate-related activities – for example, in regulatory publications, including on regulatory websites. A case in point is the publication of information on firms that participate in voluntary climate programs operated by governmental regulators. The EPA, for instance, offers several voluntary programs for the fossil-fuel industry, such as the Methane Challenge program (Section 2.2) and the Natural Gas STAR Program (Hsueh, 2020), and publicizes the names and details of participating companies (including their website addresses), as well as the details of each company's voluntary commitments.

The Israeli "red list" also constitutes a form of climate faming, as it publishes information on firms' participation in voluntary climate programs. Another example is a voluntary program for corporations run by Spain's Ministry for the Ecological Transition and the Demographic Challenge, which incorporates a greenhouse gas registry, reduction commitments, and certificates. Faming tactics are also employed by the Israeli Securities Authority, which recently began publicizing ESG reports submitted voluntarily by publicly traded firms, at the invitation of the Authority, on a dedicated webpage on its website.[25] The Authority also congratulates on its social media accounts named firms that submit ESG reports.

3.2 Typology of Regulatory Climate-Shaming Schemes

Based on the previous section, **Table 1** offers a brief overview of prominent RCS schemes from surveyed jurisdictions, arranged according to categories discussed in Section 3.1. The table includes a short policy description for each example, and an indication of the relevant jurisdiction, the year of commencement (examples in each category are listed from the newest to the oldest),[26] the goals or effects of the publication (other than shaming), the structuring of the information, the media platforms used for dissemination, the legal framework of the shaming scheme,[27] and the type of norm that serves as the basis for shaming (mandatory/voluntary norms). More information on these schemes can be found in **Appendix A** online.[28]

---

[25] www.isa.gov.il/sites/ISAEng/1489/1511/Pages/The-ISA-calls.aspx.

[26] Referring to the date on which the legal norm took effect. However, some schemes have not yet been implemented in practice, pending detailed regulations such as labeling or reporting formats.

[27] While most schemes are anchored in explicit legal powers, some are based on a broad interpretation of authority. See also Appendix A at www.cambridge.org/yadin_resources.

[28] www.cambridge.org/yadin_resources.

Table 1 Regulatory climate-shaming schemes

Scheme	Examples — Policy	Jurisdiction	Year	Goals/Effects	Structure of information	Media platform	Legal framework for publication of information	Basis of climate norm	Scheme type
Publication of enforcement actions	Publication of civil penalties imposed for breaches of climate laws and regulations	UK Environment Agency	2014	Transparency; legal certainty	Spreadsheet	Governmental website	Agency regulations	Mandatory	Soft compliance shaming
	Publication of civil penalties imposed for violations of climate agreements	UK Environment Agency	2012	Transparency; legal certainty	Spreadsheet	Governmental website	Tax regulations	Voluntary	Soft beyond-compliance shaming
	Publication of fines imposed for cap-and-trade infringements	EU member states (usually environmental ministries); UK Environment Agency	2004	Transparency; legal certainty	Lists, tables, spreadsheet	Governmental websites, national registries, official gazettes	EU directive	Mandatory	Soft compliance shaming
Rating, ranking, scoring, blacklisting	Ranking of factories	Israeli Ministry of Environmental Protection	2014	Informing; legal certainty; transparency; warning; public right to know	Top 20, Top 10, scoring, ranking, league tables, detailed reports	Ministry website, social media, press releases	Ministry regulation	Mandatory and voluntary	Soft compliance/beyond-compliance group shaming/faming
	Ranking of car models	EU member states	2001	Nudging; informing; educating	Top 10, ranking	Consumer guides, point of sale	EU directive	Voluntary	Strong beyond-compliance group shaming/faming

Climate labels								
Adverse statements in automobile advertisements	French Ministry of the Ecological Transition	2022	Nudging; educating	Statements	Company advertisements	Laws	Voluntary	Strong beyond-compliance shaming
Climate-impact ratings	Swedish Energy Agency	2021	Nudging; informing; educating	Graphic labels	On site (fuel pumps and charges in gas and charging stations)	Law and agency regulation	Voluntary	Strong beyond-compliance group shaming/faming
Climate warning labels	Municipality of Cambridge, MA	2020	Nudging; educating; informing; warning; public right to know	Warning labels	On site (fuel pumps)	City ordinance	Voluntary	Strong beyond-compliance shaming
Building energy ratings	City of New York	2019	Informing; nudging; public right to know; educating	Color-coded, textual, score-based label	On site, NYC municipal website	Local law	Voluntary	Soft beyond-compliance shaming
Vehicle carbon emissions (and fuel efficiency) grading	EU member states	2001	Nudging; educating; informing; public right to know; warning	Grading letters, statements	Point of sale, advertisements	EU directive	Voluntary	Strong beyond-compliance shaming

Table 1 (cont.)

Scheme	Examples Policy	Jurisdiction	Year	Goals/ Effects	Structure of information	Media platform	Legal framework for publication of information	Basis of climate norm	Scheme type
	Product energy labels (such as car tires) and registry	EU member states	1994- (labels), 2004- (grading), 2021- (database)	Informing; educating; nudging; warning	Color-coded grading letters	On products, database, QR code, advertisements	EU regulation	Voluntary	Soft beyond-compliance shaming/faming
Climate disclosure in company reporting	Companies' climate-related financial disclosures	UK Department for Business, Energy and Industrial Strategy	2022	Informing	Company reporting	Company websites	Department regulations	Voluntary	Soft beyond-compliance shaming/faming
	Digital companies' climate policy disclosures	France	2022	Informing; educating; nudging	Company reporting	TBD	Law	Voluntary and mandatory	Soft beyond-compliance/ compliance shaming/faming
Emission databases	Publicizing information via "Facility-Level Information on GreenHouse gases Tool" (FLIGHT)	US EPA	2021	Informing; educating; public right to know	Database (including maps, graphs, lists, and charts)	Agency website	Notices and memorandums	Voluntary	Strong beyond-compliance group shaming/faming
Climate naming and faming	Publicizing companies' ESG reports	Israel Securities Authority	2021	Informing	List	Authority website and social media accounts	Ministry of Environmental Protection regulation	Voluntary	Beyond-compliance group faming

Instrument	Agency	Year	Purpose	Format	Disclosure channel	Legal basis	Participation	Mechanism
Publicizing information on firms that joined voluntary Agency programs	US EPA	2016 (Methane Challenge Program); 1993 (Natural Gas STAR Program)	Informing	Lists with search engines	Agency website	Federal regulation	Voluntary	Beyond-compliance group faming
Publicizing information on firms that joined voluntary private and regulatory programs	Israeli Ministry of Environmental Protection	2014	Informing	Ministry reports	Ministry website	Ministry regulation	Voluntary	Beyond-compliance group faming
Carbon footprint registries	Spanish Ministry for the Ecological Transition and Demographic Challenge	2014	Informing; educating	Database, certificates	Agency website	Royal decree	Voluntary	Beyond-compliance group faming
Publication of sector and company performance relating to regulatory climate agreements	UK Environment Agency	2012	Legal certainty; transparency	Spreadsheet	Governmental website	Tax regulations	Voluntary	Beyond-compliance group faming/shaming

The table further indicates scheme types, based on the following dimensions: soft or strong shaming; mandatory or voluntary norm that serves as a basis for the publication, meaning the nature of the climate norm (compliance/beyond-compliance); group shaming; and shaming or faming. The classification of soft versus strong shaming was based on the prominence of the shaming element relative to other goals or effects presented in **Figure 1** and accompanying text, like informing or warning. These appear in the table under goals/effects. Other parameters, including structure of information and media platform, were also taken into account in this context. For example, short ranking lists prominently displayed in points of purchase were considered more shaming than inaccessible Excel files posted on an agency's website. The typology also takes into account the level of focus on climate relative to other issues, such as environment. The more focused they are on climate, the more the surveyed schemes were regarded as stronger forms of RCS. All faming schemes are regarded as inherently soft modes of shaming, and so the soft/strong parameter was only indicated for shaming schemes.

The classification of compliance/beyond-compliance was created using the "basis of climate norm." Generally, the legal framework for shaming schemes relates to laws, rules, regulations, and the like that relate to the publication of the shaming information. Conversely, the "basis of climate norm" relates to the legal status of the specific corporate behavior that is targeted by these shaming tools. For example, climate labels can be anchored in laws and regulations that govern their design and set other obligations on companies subject to labeling requirements; at the same time, they are often based on a voluntary climate norm in the sense that firms do not have to improve climate performance or meet a climate-related standard. The behavior that is targeted by RS in these cases is not a violation of the law. I have referred to this in the table, under "scheme type," as either "compliance" (relating to mandatory base norms) or "beyond-compliance" (relating to voluntary base norms).

"Group shaming" (or faming) relates to instances in which stakeholders can easily and immediately compare companies' performance based on the initial publication, without having to take any additional action. Posting lists ranking companies on social media was therefore regarded as group shaming, while company reports, which are hard to compare, were not. It should also be noted that group shaming relates to the publication of company-specific information and not to the shaming of entire sectors.

Based on these dimensions, the following typologies emerge:

| Soft compliance shaming |
| Soft beyond-compliance shaming |
| Soft beyond-compliance shaming/faming |
| Soft beyond-compliance/compliance shaming/faming |
| Soft compliance/beyond-compliance group shaming/faming* |
| Beyond-compliance group faming |
| Beyond-compliance group faming/shaming |
| Strong beyond-compliance shaming |
| Strong beyond-compliance group shaming/faming |

*Refers to ranking and rating schemes which are based on both mandatory and voluntary standards and can also shame or fame companies, depending on companies' ranking or rating.

Figure 4 A typology of regulatory climate-shaming schemes

It should be noted that classifications such as strong or soft shaming, and group shaming, are not always definite, and may vary depending on different design factors. Additionally, some dimensions can be placed on a scale rather than presented as binary (e.g. highly, moderately, or very mildly shaming). The typologies presented in **Figure 4** are meant to aid in developing a conceptual, theoretical, and policy-oriented discussion on RCS, rather than establishing strict categories for their own intrinsic value. Additional typologies can be developed in theory (see **Appendix B**),[29] or based on more examples from more jurisdictions.

---

[29] www.cambridge.org/yadin_resources.

4 Policy Directions for Regulatory Climate Shaming

Building on the previous sections, this section suggests five main interrelated directions for designing regulatory climate-shaming policies, namely: using stronger shaming messages; communicating climate risks; emphasizing corporate injustices; separating climate shaming from eco-shaming; and making publications more publicly accessible. **Appendix B** online[30] includes examples of recommended schemes for RCS, based on the principles developed in this section.

Discussing the design of RCS policies is especially timely, for several reasons: first, many regulatory climate-shaming schemes surveyed in Section 3 have been developed and implemented very recently, mostly in the past year or two (see **Table 1**), and many more are currently being considered; second, RCS schemes that date further back have recently been updated and enhanced with new features; and third, there is growing public and regulatory awareness of the urgency and seriousness of climate change, especially in light of newly published IPCC reports (Bell et al., 2021; Hughes et al., 2020; Leiserowitz et al., 2021a). These changes also render the limited scholarship on climate information-sharing tools largely outdated.

Information-Sharing Schemes Should Become More Shaming

One of the most distinctive features of RS is the use of condemning, disapproving, and criticizing statements that focus on the moral aspects of corporate activities (see Section 2.1). Accordingly, "shaming" was defined for the purpose of this Element as an action that intends to inflict reputational harm (Section 1.3). In the same vein, the discussion of the differences between various forms of information-sharing policies (in Section 2) underscored that RS is premised on elements of condemnation, public opinion, and reputational risks. The related element of inviting relevant stakeholders to also shame companies or actively respond to the shaming information was also discussed in Section 2 as integral to the concept of RS. Indeed, "shaming" companies into better climate performance serves as a central notion of RCS.

However, while some of the climate-shaming examples that were studied in this Element can be considered forms of "strong shaming," many others were categorized as "soft shaming" schemes (**Table 1**). Section 3 has demonstrated that strong climate shaming is executed, for example, by rankings, mandatory announcements in commercials which recommend using the product less, and mandatory labeling on products which emphasizes the dangers inherent in the

---

[30] www.cambridge.org/yadin_resources.

product. Soft forms of climate shaming are those that only indirectly shame; are inaccessible in terms of form or content (e.g. via company reports); concentrate on faming; focus on goals such as informing and nudging, for example, consumers; or emphasize issues other than climate. This part of the section argues that in order to fully realize the potential of RCS to curb climate change, shaming schemes need to be more direct, judgmental, and critical toward firms' contributions to the exacerbation of climate change. In effect, all following parts in this section, while separate and distinct, also support this central idea.[31]

Research in the fields of environmental and climate disclosure, as well as RS, underscore the importance of the shaming dimension in informational policies. For example, Van Erp (2011) demonstrates the failure of disclosure policies that "name without shame," and Shimshack (2020) argues that out of all the more common informational policies, the most consistently effective environmental disclosure policy is "naming and shaming." Similarly, scholars argue that the success of various information-sharing policies derives from their shaming-like effect – namely, their ability to communicate messages that drive stakeholders to pressure firms to improve performance and compliance (Bonetti et al., 2023; Johnson, 2020; Van Erp, 2010). In this vein, Fung and O'Rourke (2000) argue that while the EPA's TRI was not initially created with the aim of leveraging public pressure to drive companies to reduce chemical releases (via a "shaming" mechanism), this was in effect the main reason for its success.

More specifically, based on the TRI's success, Fung and O'Rourke (2000) have argued that environmental policymakers should develop similar informational schemes that drive negative attention to companies, thereby enabling stakeholders to exert pressure on those firms. These and similar studies suggest that simpler, more explicit shaming messages could be effective in inducing reactions from stakeholders and motivating corporations to adjust their business practices. For instance, Fung and O'Rourke (2000) contend that when technical data posted online by the EPA was presented by the media and environmental organizations in a more shaming form, for example, blacklists that highlight health risks posed by the worst offenders, it affected corporate behavior more significantly.

---

[31] Even if regulators do not frame these practices as shaming, or do not intend for the information-sharing practices to act as an enforcement measure (though for most surveyed policies this does not seem to be the case), my argument in this section should be read as a call for regulators to develop regulatory climate-shaming tools in accordance with the recommendations put forward.

Indeed, RS schemes with a strong shaming component (generally one which also condemns or criticizes corporate behavior) are being used or have been implemented with great success in various environmental and public health areas discussed in this Element. These include OSHA's policy of shaming noncompliant employers on social media and via press releases, with company-specific condemnatory statements; the labeling of poorly performing factories with black stars and excelling factories with gold stars in Indonesia; the top-ten environmental lists in the Netherlands; and the publication of blacklists based on the TRI.

This approach of using strong shaming schemes to achieve environmental goals has recently been taking hold in various other forms and jurisdictions as well. For example, the Israeli Ministry of Environmental Protection began posting on social media, mainly on Facebook, the names of companies that are infringing on environmental laws and regulations.[32] Noticeably, these posts are often accompanied with condemnatory statements by the Ministry, specifying the ways in which the company is endangering the public and the environment, and underscoring the seriousness of the offenses and infringements. Some posts emphasize the company's history of violations and its unsatisfactory approach to environmental compliance, adding to the moral condemnation inherent in the message. Some posts also include pictures of the shamed facilities, such as a natural gas drilling rig or a factory yard filled with hazardous materials.

Another highly shaming environmental approach is taken by the Irish EPA, which scores and ranks facilities according to environmental compliance and performance, including complaints made against them.[33] The list is not only passively published on the agency's website but also circulated as a press release, highlighting companies that received the highest number of complaints or that have not improved their scores over time. Press releases are typically accompanied with condemnatory regulatory statements, for example, stating that it is unacceptable that certain industrial and waste sites caused great distress to those living around them by emitting offensive and potentially harmful odors, especially during the COVID-19 pandemic.

It is suggested that regulators adopt similar policy directions for climate-change informational schemes as well. A more shaming approach in the climate context can also serve stakeholders who are interested in becoming more active. Recent surveys conducted in the United States and elsewhere have indicated that the public seeks clear direction from regulators as to which companies should be "punished" for unsatisfactory climate policies – for example, via consumer activism (see Sections 1 and 2). This need can be

---

[32] See note 22.
[33] See https://www.epa.ie/our-services/compliance–enforcement/whats-happening/national-prior ity-sites-list.

better fulfilled by clearer shaming messages that spotlight misbehaving firms and explain why and how these firms have acted wrongly. A more salient shaming feature in climate disclosure policies could also drive shaming targets to improve performance regardless of stakeholders' actual responses. This is because businesses tend to have exaggerated expectations regarding stakeholders' responses to disclosed negative information, for example, via energy efficiency labeling (Loewenstein et al., 2014).

One way to make shaming schemes more strongly shaming is to include more condemning and criticizing messages and statements in publications (the next parts of this section present other ways to achieve strong shaming). For instance, the social and environmental license of a firm and the gravity of the harm that it has caused could play a key role in the shaming message. Such a shaming message could say, for example, that named and shamed companies are not doing enough to curb greenhouse gas emissions or are acting in a way that exacerbates climate change, and that climate change is risking the lives of people. This direction is further developed in the following parts, which discuss communicating climate risks and firms' indirect contribution to climate change via climate obstruction.

A stronger shaming approach also means that shaming schemes need to move from focusing on informing consumers and investors, to utilizing various other stakeholders to exert pressure on firms. As detailed in Section 3 and **Appendix A**,[34] many European and American schemes tend to focus on nudging consumers and investors toward climate-friendly choices and on providing them with useful information to support decision-making (indicated as "nudging" and "informing" goals in **Table 1** and in **Figure 1**), rather than on exposing corporate responsibility for climate change. This approach overlooks the potential to engage additional relevant stakeholders like employees, NGOs, the media, and business actors, who were successfully harnessed in other, related areas, such as public health and environmental protection, using a regulation-by-shaming approach (see Section 2.1). Recent studies discussed in Section 2.2 further point to the ability of various stakeholders to take part in the climate-shaming process and drive corporate action, including suppliers, retailers, consumers, shareholders, transporters, and producers. It is therefore recommended that climate-shaming schemes developed by regulators and legislators are not limited to appealing mainly to consumers and shareholders.

Relatedly, regulators usually refer to schemes discussed in this Element as transparency, disclosure, or informational (supporting decision-making) tools (**Appendix A**[35] and Section 2.1), thereby neglecting to emphasize

---

[34] www.cambridge.org/yadin_resources.
[35] www.cambridge.org/yadin_resources.

and develop the potential of climate shaming as a tool that invites the public to shame companies into compliance while also strengthening social climate norms. Indeed, people and policymakers are currently too focused on individual climate responsibility and on making behavioral changes at an individual level, while insufficient attention is given to the central role of corporations in the climate crisis (Jacquet, 2015). While informing stakeholders is certainly a worthy goal, RCS policies could achieve more by explicitly and declaredly shaming companies. Although strong shaming schemes may prove to be more challenging than soft shaming schemes from legal and political points of view, the circumstances of this grave crisis we are currently facing may provide policymakers with the required public legitimacy and support, in line with the targeted transparency theory discussed in this Element.

Informational climate policies could also become more shaming by expanding the current target sectors, which in surveyed jurisdictions are mainly industrial facilities with high carbon footprint, the automobile industry, and energy and fuel (see **Table 1** and Section 3.1). Regulators should continue expanding their shaming policies to target diverse industries and communicate the industries' responsibility for climate change – in accordance with the policy directions discussed in Section 3.1, relating to new climate-shaming schemes currently being considered by diverse regulators in various fields. Building on theories suggesting that information-sharing policies can be more effective when the information published is new and surprising (Stephan, 2002), the public may be surprised to learn that many companies in sectors such as fashion, advertising, finance, and food are also contributing to climate change, and this new information may make them more willing to exert pressure.

Another element of increasing shaming (inflicting greater corporate reputational harm) relates to faming other firms. Generally, it has been suggested that based on prospect theory, which contends that people are more sensitive to loss than gain, naming and shaming is a more powerful and effective measure than naming and faming, in the context of regulation of firms (Bevan & Wilson, 2013). In this vein, scholars have recommended focusing negative attention on products and companies that create the highest risks – for example, by only labeling products that have a significant carbon footprint (Cohen & Viscusi, 2012) – and concentrating on the worst environmental performers (Fung & O'Rourke, 2000). However, faming can still play an important role in RCS policies.

Studies surveyed in Sections 1 and 2 demonstrated consumers' willingness to support firms based on their climate policies, in addition to their willingness to punish nonclimate-friendly firms. Section 2 also discussed the potential of climate faming for providing firms with the opportunity to showcase their ethical behavior and for encouraging institutional isomorphism, citing research in the field of targeted environmental transparency (Fung et al., 2007). Thus, theoretically, informational climate policies could become strong shaming schemes by combining shaming and faming in a single publication, using a group shaming/faming scheme type. For example, the use of rankings, in which stakeholders can easily compare high-performing and low-performing companies based on their climate performance, can further highlight good actors' achievements by presenting them next to the worst performers, and also shame bad actors more by presenting them next to excelling companies.

Such schemes can certainly be based on legal infringements and compliance, but should focus more on beyond-compliance corporate actions and performance, which are more relevant in the climate context because climate law and regulation are generally underdeveloped (Section 1.1). Beyond-compliance shaming has also been shown to be effective by studies of environmental and climate disclosure policies discussed in this Element (Section 3). Strong compliance/beyond-compliance group shaming/faming schemes are therefore a recommended policy direction for climate-change regulation (an example is provided in **Appendix B**).[36]

Schemes such as OSHA's press releases and social media publications, which individually condemn infringing firms, may also prove successful in the climate context. These schemes are arguably more shaming than group shaming schemes, and studies have shown them to be effective in fields such as occupational safety. However, the legal infrastructure of occupational safety regulation is generally much more established than the legal infrastructure of climate-change regulation, in terms of hard law (command-and-control) – though occupational safety regulation based on command-and-control is itself often flawed and limited. By contrast, climate law and regulation is virtually nonexistent or highly nascent in some jurisdictions (Section 1.1). Therefore, RCS schemes based on shaming individual firms (rather than "group shaming") may need to be more innovative and to be grounded in soft law and "beyond-compliance" approaches rather than in legal compliance with mandatory laws and regulations.

Shaming Schemes Should Explain Climate Risks

In many countries, including the United States, climate change became a public issue back in the late 1980s (Dessler & Parson, 2019). Since then, public awareness of climate change has grown steadily (Sections 1 and 2).

---

[36] www.cambridge.org/yadin_resources.

However, a great deal of confusion still exists worldwide (Eichhorn et al., 2020). Indeed, climate change is for most people a more complicated and intangible topic than environmental or public health issues (Section 1). In addition, public understanding of climate change is hampered by a great deal of misinformation and disinformation that result from corporate climate-obstruction tactics such as climate denial and climate washing (Section 2.2).

Yet a look at existing regulatory climate-shaming schemes (Section 3) reveals that, while some schemes do explain the connection between a company, or its products, and climate change, as well as the meaning of climate change, others do not. It is thus recommended that climate-shaming schemes are accompanied by brief, clear, accessible, and communicative messages, which do not burden potential stakeholders with too much technical and scientific information (Fung & O'Rourke, 2000; Fung et al., 2007) but give them sufficient context and understanding to help them effectively carry out their roles as private enforcers of climate-change norms.

In addition to the importance of communicating understandable information, the theory of RS maintains that it is easier to shame using new and shocking information, especially when it relates to grave risks such as those involving pollution and other environmental hazards (Stephan, 2002). Even when the information merely reveals that corporate action, like environmental pollution, creates more risk than was initially understood by stakeholders, it can drive community members, interest groups, investors, and the media to act (Stephan, 2002). Therefore, in order to become more effective, just, and fair, climate-shaming schemes need to help people understand the grave risks created by firms and posed by climate change, as well as the real nature, causes, and impacts of climate change (Section 1.1).

Given the need to utilize new, risk-related information to grab the public's attention (Stephan, 2002), such information should focus on the relative newness of the knowledge that has been accumulated and established on the acuteness of climate change and its alarming progression and implications. Namely, the increasingly alarming picture painted by each new IPCC report – relating to the continuing rises in the earth's temperature and the very high risk posed to mankind and the entire planet – should be utilized in shaming activities.

The idea is not to intentionally provoke people, but to make shaming mechanisms more effective by being more understandable and conspicuous. This policy direction also supports the goal discussed in the previous part, of making publications more shaming, and is also in line with safeguarding the public's right to know and ensuring its ability to make informed choices.

In addition, communicating climate risks more clearly may help RS schemes gain more public legitimacy – an important challenge discussed in this Element. The information relating to climate risks must also be regularly verified and updated so as to not lose credibility (Shimshack, 2020), thereby also guarding against some of the risks discussed in Section 2 involving the image and credibility of shaming regulators. Frequent updates can also reduce the ability of firms to behave strategically and evade meaningful compliance with climate norms (see Section 2) by achieving irrelevant or vague standards (Cohen & Viscusi, 2012).

Of course, the effectiveness of providing more information on climate risks may be somewhat weaker when considering "informational avoidance" among stakeholders – the tendency to intentionally ignore fear-inducing or uncomfortable information (Loewenstein et al., 2014; Reisch et al., 2021; Sunstein, 2020). The climate crisis evokes adverse feelings of guilt, shame, anxiety, and fear (Fredericks, 2021), which people may prefer to avoid, and thus providing more information on the implications and effects of climate change on their future may exacerbate unpleasant feelings and nudge people to intentionally disregard such information (Sunstein, 2020). However, theory also suggests that when information is perceived as important, people can overcome informational avoidance (Sunstein, 2020). And indeed, most people in many countries now regard the issue of climate change as highly important (Section 1.2).

Shaming Schemes Should Explain Companies' Indirect Involvement in Climate Change

While currently highly focused on exposing direct corporate involvement in climate change via emissions (see Section 3 and **Table 1**), RCS should also expose companies' indirect involvement in the climate crisis. As discussed above, the theory of RS maintains that it is easier to shame using information that is shocking, new, or creates great surprise relating to the form or extent of firms' adverse behavior, especially when it involves grave harms and risks (Stephan, 2002). In fact, "shock-and-shame cycle" theory suggests that the greater the shock created by information-based environmental schemes, the better the chance of the policy motivating people to act. It has also been argued that corporations suffer substantial reputational losses especially when they are exposed as lying to, deceiving, and defrauding stakeholders such as investors, customers, consumers, and suppliers (Karpoff, 2012). Research further suggests that the motivation to act in response to environmental pollution data may be especially high when the information revealed appeals to people's sense of injustice (Stephan, 2002). In this vein, the information used by regulators to climate shame should concentrate on the real extent of the responsibility of firms for climate change, which may shock stakeholders and drive them to action by enhancing injustices caused by firms. Specifically, RCS should expose corporate climate-obstruction tactics discussed in this Element.

As explored in Section 2.2, climate-obstruction activities were carried out for decades by the fossil-fuel industry, which for the sake of profit risked the lives of millions, actively misled policymakers and the public, and lobbied against climate laws and regulations which could have slowed climate change. The fossil-fuel industry and other carbon-intensive and supporting industries are still engaging in climate-denial and climate-washing practices that exacerbate climate change (Freese, 2020; Michaels, 2008, 2020; Supran & Oreskes, 2021). Importantly, these corporate actions are deceitful not only toward regulators and other policymakers but also toward stakeholders, including investors, consumers, residents, employees, and the public at large, causing them great injustice.

Indeed, the idea that climate shaming is difficult to implement successfully because it relates to diffused externalities (Cohen & Viscusi, 2012) and does not directly affect people in the same way as public safety, health and environmental pollution (see Section 1.2) seems to be losing traction. A wave of litigation now being brought against companies, especially in the fossil-fuel sector, usefully illustrates the specific harms caused to specific stakeholders due to climate obstruction. These claims allege that companies have misled consumers, investors, and residents by downplaying the grave implications of climate change and their own role in causing it (Yadin, forthcoming-b). Indeed, climate obstruction directly affects various stakeholders.

Specifically, firms' climate-washing and climate-denial practices directly manipulate stakeholders, who are denied the ability to autonomously decide which products to consume and with which companies to work. Such practices can also inflict "identity harms" on stakeholders like consumers, employees, and suppliers (Dadush, 2018) – a term referring to the anguish experienced by those who learn that their efforts to act in ways that match their personal values have been undermined by corporate deception. Residents are also affected by corporate climate denial and climate washing, especially in areas in which costly adaptation efforts which are needed to secure safe living conditions were not taken in time by cities and states, due to misleading presentations by the fossil-fuel sector about the safety of its products and activities. Investors have also been deceived by companies via financial projections supplied by firms as to their climate risks and their exposure to expected climate-related losses. Of course, the public at large, and especially young people, are already affected and expected to be affected even further in the coming years by the devastating implications of climate change, which are further exacerbated by climate obstruction (Section 1.1).

In addition to these direct harms via manipulation, misinformation, and concealment, climate-obstruction tactics wrong stakeholders because they are employed by firms specifically in order to silence public activism and suppress meaningful public discourse relating to energy policies (Yadin, forthcoming-b). In designing climate-shaming schemes, regulators should therefore make use of information concerning the intentional deceit inherent in climate obstruction, and not only of information on direct emissions and regulatory violations of climate rules and regulations.

Moreover, since the phenomenon of climate obstruction is not common knowledge (McGreal, 2021), shaming relating to corporate climate obstruction may prove especially effective by offering important new information to stakeholders who were previously unaware of the mechanisms and extent of climate obstruction. While portions of the public are generally familiar with issues such as greenhouse gas emissions (usually associated with oil and gas companies), most people are less familiar with companies' contribution to the delaying and obstruction of climate mitigation (McGreal, 2021).

Importantly, regulators should focus on condemning behaviors that corporations can change, for instance in areas in which emissions can be reduced by using available technologies (Cohen & Viscusi, 2012; Delmas et al., 2010). Practices of climate obstruction are a good basis for shaming in this respect because they can (and should) always be eliminated. Additionally, exposing them has a stronger shaming effect, revealing a fuller picture of firms' contribution to climate change and exposing corporate manipulation and injustice toward stakeholders, and can thus drive action by both firms and stakeholders. This direction could also enhance the legitimacy of shaming policies by better explaining why shaming is necessary, and strengthen climate-change norms. In summation, focusing on climate obstruction can serve to enhance shaming policies' effectiveness, fairness, and legitimacy.

In order to carry out shaming schemes based on companies' indirect involvement in climate change, regulators could blacklist companies suspected of circumventing or obstructing climate regulation, being involved in climate denial, or engaging in climate washing, using a strong beyond-compliance group shaming scheme type (an example is provided in **Appendix B**).[37] Using

---

[37] www.cambridge.org/yadin_resources.

this scheme type, or closely related schemes (e.g. schemes that do not use group shaming, or that are based on legal infringements), regulators could expose any kind of corporate misrepresentation or deceit. This could include publicizing infringements of climate regulation agreements, as in the UK approach; uncovering selective, misleading, or incomplete disclosures in companies' voluntary or mandatory reports, statements, and labels; and publicizing instances of shirking within voluntary programs (Taebi & Safari, 2017). An accompanying statement as to the magnitude and meaning of corporate climate obstruction and to the role that corporations have played and are still playing in bringing about and exacerbating this crisis, not only directly but also indirectly, may prove useful in this context.

Similar tactics, focusing on blacklisting companies that are manipulating the market by circumventing regulatory or ethical requirements, are now being employed or considered by regulators like the ECB and the FDA (Sections 2.1 and 3.1). Another relevant example is the UK Competition and Markets Authority (CMA), which plans to name and shame greenwashing fashion companies (Horton, 2022).

Climate litigation, which in a way is already being harnessed by regulators who name and shame companies for climate infringements, can also be utilized in the context of exposing companies' climate-denial practices. In light of the general scarcity of climate laws and regulations, climate suits discussed in this part are usually based on innovative legal doctrines relating to climate denial (Yadin, forthcoming-b). In these cases, the firms' formal responses often consist of more climate-denial arguments (Yadin, forthcoming-b). Regulators can thus publicize lists of ongoing climate-denial litigation brought against firms, accompanied by summaries of arguments from both sides, to maximize the shaming effect of climate litigation (discussed in Section 1.2) and further shame firms into implementing climate-friendly mitigation and adaptation measures (and possibly into settling lawsuits).

Climate Shaming Should Be Separated from Eco-Shaming

Examples reviewed in Section 3 show that in some cases, climate-shaming schemes are integrated within eco-shaming schemes. For example, companies and products are sometimes scored, ranked, and labeled based on combined environmental and climate performance (Cohen & Viscusi, 2012; Taufique et al., 2022). Climate issues do not necessarily take the lead in such integrated policies, leaving the climate-shaming aspect of the information relatively inconspicuous. For instance, the Israeli "red list" (Section 3.1) ranks companies based mainly on their environmental risk and performance, and only marginally on their adoption of voluntary climate standards.

Generally, while climate shaming focuses on issues such as greenhouse gas emissions, and the carbon footprint of products, services, and companies, eco-shaming focuses on such issues as air, water, and soil pollution, waste and hazardous material management, and odor nuisance. Importantly, as explained in Section 1.1, climate change involves much more than just changes in weather patterns, and it is not simply another environmental issue, although it is often mistakenly framed and regarded as such (Egan et al., 2022). Unlike conventional environmental issues, climate change is an urgent, global problem that poses a real threat to humankind and should therefore receive appropriate and distinct attention from policymakers. Shaming schemes in which environmental and climate-related information are intertwined can not only dilute the climate-shaming message but also send the wrong signal to stakeholders. Conversely, by forming independent climate-shaming schemes, regulators can signal the importance of the issue to firms and stakeholders.

Additionally, mixing environmental and climate-related information may cause confusion to relevant stakeholders. Research on informational policies discussed in Section 2 (under frameworks such as RS, disclosure regulation, or targeted transparency) has stressed the importance of relaying clear messages and understandable information to the public. Indeed, this principle is also valid in the environmental and climate context, and concentrating on more than one issue can create informational overload (Cohen & Viscusi, 2012; Downar et al., 2021; Fung & O'Rourke, 2000). Of course, RCS that does not accurately and clearly reflect the company's level of commitment to climate mitigation will not be able to effectively harness stakeholders to the shaming process.

From a related perspective, regulatory climate-shaming and eco-shaming methods should be carried out separately because they relate to different problems that have different manifestations, and therefore require different tool designs. For example, stakeholders possess different levels of comprehension of the causes and impacts of climate change than of corporate activities that affect environmental protection (see Section 1). Namely, while environmental noncompliance is often very tangible – manifesting for example, in changes in odors and visibility, or creating breathing difficulties, in cases of air pollution; or in changes in the color, texture, and smell of water or soil – the dangers of climate change are undetectable to human senses. Climate change is also less immediate in its effects on living creatures than are environmental incidents. Consider, for example, major oil spills in the ocean, which immediately harm marine animals. The causes, mechanisms, and impacts of climate change are harder to detect and understand and are less intuitive, warranting, for example, the use of explanatory messages discussed in this section.

Scholarship discussed in Section 2 regarding companies' tendencies and abilities to manipulate disclosure schemes to avoid reputational harm, while also misleading stakeholders (and sometimes regulators), can also be of relevance here. Mixing environmental and climate data could generate misleading scoring, rating, and labeling in cases where firms are performing well environmentally but are substantially deficient in cutting their greenhouse gas emissions, for example. In other words, separating the two issues can help safeguard against instances of climate washing in which firms that are underperforming on the climate-change front attain good regulatory scores and ratings resulting from corporate manipulation and/or creative compliance (see Section 2.1).

In the same vein, the separation of environmental and climate-related information can improve the methods for calculating corporate contribution to climate change (e.g. differentiating between scope 1, 2, and 3 emissions, which relate to direct and indirect emissions, including from supply chains). If RCS is developed as a separate, independent tool from regulatory eco-shaming, it can become more professional, advanced, and accurate. This, in turn, could also help avoid (even inadvertent) forms of greenwashing, climate washing, or otherwise misleading statements, grades, or rankings, which are often a pitfall for implementing disclosure-based policies (see Section 2.1).

In addition, eco-shaming and climate-shaming tools are legally different: as discussed in Sections 1.1 and 2.2, by and large, countries have not yet developed a thick legal basis for climate-related corporate obligations. Conversely, environmental law and regulation is highly developed in many nations. Therefore, it is not surprising that many climate-shaming schemes are based on "beyond-compliance" norms like regulatory agreements and voluntary programs. Certainly, many countries are currently in the process of legislating climate laws and regulations (Section 1.1), but these are usually separated from extant environmental laws, and contain a different set of rules and principles. Relatedly, the justifications and rationales for environmental regulation, including RS, are different from the justifications for climate-change regulation, including by shaming (see Section 2.2). These differences warrant separate legal arrangements for regulatory eco-shaming and RCS. It would therefore be easier and would make more sense to separate these tools for legal purposes.

For example, it has been suggested (Yadin, forthcoming-a) that regulatory eco-shaming schemes be subject to several administrative safeguards, such as conducting hearings, giving warnings, producing guidelines, and implementing privacy procedures (see Section 2.1). These procedural measures aim to ensure fairness toward shamed firms while balancing environmental regulation effectiveness. However, it is plausible that when it comes to emergencies such as the climate crisis, administrative procedural measures can be less stringent (Daly, 2021: 78).

Climate-Shaming Schemes Need to Become More Accessible

Some climate-shaming schemes discussed in Section 3 were characterized as "soft shaming" due to their inaccessibility to relevant stakeholders. For instance, the UK Environment Agency, which names and shames companies by publicizing their detailed infringements of climate law and regulation, does so by means of Excel files located in its website's inner pages, making the information not easily accessible to the public. Similarly, many EU member states publicize cap-and-trade infringements in an inaccessible manner (see Section 3). Mandatory climate disclosures in companies' financial reports and via emissions databases also tend to be less accessible, because they contain much unprocessed information that requires time, effort, and expertise to parse and understand.

Research discussed in this Element (Section 2.1) on the effectiveness of various naming-and-shaming, "targeted disclosure," and other closely related instruments has underscored the importance of clear, usable, and accessible information. Studies have indicated, for example, the importance of designing clear and short forms of information sharing, rather than publishing all-inclusive databases with a lot of technical information (Cohen & Viscusi, 2012; Downar et al., 2021; Fung & O'Rourke, 2000; Van Erp, 2010). Researchers have also pointed to the importance of shaping and delivering information to stakeholders in a manner that is not too costly to obtain in terms of time and resources, for example, by putting up hygiene inspection grades on restaurant windows (Fung et al., 2007: 55). In the context of climate change, Downar et al. (2021) have noted that significant greenhouse gas reductions took place only when UK regulators provided stakeholders with company-specific carbon footprint information via a searchable emissions database.

Against this background, it is recommended that regulators disseminate processed and succinct climate-shaming information, rather than relying on large quantities of inaccessible technical data on greenhouse gas emissions and climate infringements and making it only passively available to the public on regulatory websites. As climate-reporting obligations for companies – for example, via emission databases or company reports – proliferate in various jurisdictions, regulators should also build on these datasets to publicize shaming information to the public in a more accessible manner, such as in the form of top-10 lists.

In order to increase accessibility to shaming information, regulators should also utilize various media outlets and new digital capabilities, focusing on social media, apps, interactive online tools, and infographics, which can actively publicize and spread information in easy-to-digest formats.

The following recommendations are therefore offered to policymakers designing, implementing, and adjusting RCS schemes (see **Figure 5**).

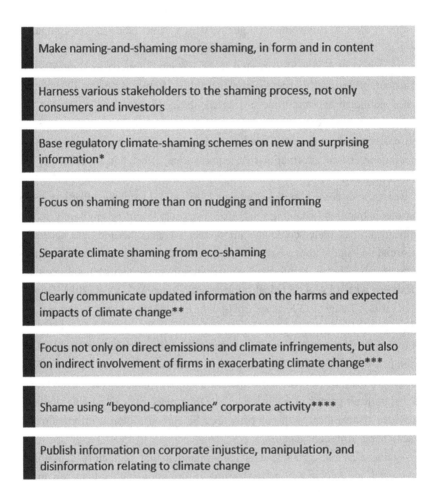

Make naming-and-shaming more shaming, in form and in content

Harness various stakeholders to the shaming process, not only consumers and investors

Base regulatory climate-shaming schemes on new and surprising information*

Focus on shaming more than on nudging and informing

Separate climate shaming from eco-shaming

Clearly communicate updated information on the harms and expected impacts of climate change**

Focus not only on direct emissions and climate infringements, but also on indirect involvement of firms in exacerbating climate change***

Shame using "beyond-compliance" corporate activity****

Publish information on corporate injustice, manipulation, and disinformation relating to climate change

* For example, highlighting various industries that contribute to climate change, or the practice of corporate climate obstruction.

** Without burdening relevant stakeholders with overly complicated information.

*** For example, via climate denial and climate washing.

**** Such as shirking and infringements relating to regulatory agreements and voluntary programs.

Figure 5 Policy recommendations

5 Conclusion

This Element has put forward a descriptive and normative theory for regulatory climate shaming, arguing that climate shaming is both a desirable and feasible approach and one already in use and under development in various forms and jurisdictions worldwide. On a descriptive level, I have surveyed regulatory laws, rules, regulations, local laws, and governmental information-sharing practices that facilitate or aim to facilitate mechanisms of climate-shaming corporations – in EU member states, the United States, the United Kingdom, and Israel (Section 3). The rich array of RCS tools that emerged from this review was used to construct a typology of climate-shaming tools. This typology was based on various tool categories and on dimensions like soft or strong shaming, the nature of the climate norm, group shaming, and shaming or faming. Shaming tools discussed in the Element include the publication of legal sanctions; rating and ranking schemes; the use of labels, registries, and databases; company disclosure obligations; and positive publications.

On a normative level, the Element has sought to demonstrate that the shaming of firms by regulators is a necessary and justified approach in the fight against climate change. Namely, I have argued that RCS is needed in light of the obvious failures of current national and international climate law, regulation, and governance, and the gravity of the climate crisis; that the climate crisis is a suitable topic for regulatory shaming; and that RCS could both enjoy legitimacy and prove effective and successful – especially due to the public's growing concern and willingness to act, and to the features of the underlying problem of corporate climate obstruction.

An examination of the research and practice of governmental information-sharing policies in climate and climate-related fields, such as environmental and public health regulation, highlighted the advantages and limitations of shaming as a tool to encourage corporate climate compliance (Section 2). Indeed, information-based policies are complex and should not be guided by a simplistic assumption that all forms of disclosure, transparency, nudge, or naming and shaming are always desired (Ben-Shahar & Schneider, 2014; Fung et al., 2007; Shimshack, 2020; Sunstein, 2020; Van Erp, 2021). A careful multidisciplinary analysis of RS studies and environmental policy scholarship applied to the unique context of the climate crisis was used to suggest policy directions for RCS as it moves forward (Section 4).

The RCS framework developed in this Element is meant to deal with the relative sluggishness of climate law and regulation worldwide (Section 1) by moving from hard to soft law and soft regulation; from legal license to operate to social license to operate; and from conventional legal sanctioning to arguably less conventional reputational sanctioning. Importantly, RS is not intended to inflict emotional harm, but only to cause reputational and financial harms to artificial legal entities (such as companies) as a regulatory tool with the regulatory purpose of slowing climate change. In this way, it is meant to pressure firms into doing the right thing and adopting climate-friendly strategies, for fear of financial repercussions.

It is also important to note that RCS is not merely disclosure nor transparency. Rather, it is a unique regulatory tool that actively targets the attention and sentiment of the public, and utilizes corporate reputational sensitivities relating to climate change. It is a sophisticated, modern approach to regulation that, as I have argued, could provide an effective response to some of the most problematic aspects of climate-change policy today, including those relating to climate-denial and climate-washing practices employed by firms. By moving the regulatory battlefield to more amenable and effective arenas – those of public opinion, information, and communication – I contended that regulators have a real chance to achieve faster progress in mitigating climate change and nudging corporate action.

Though some may find it a provocative or aggressive tool, now is the time to take climate regulation to the next level and implement RCS. Each new IPCC report escalates warnings of a catastrophe on a global scale, consistently pointing to the intensifying urgency of the crisis and the fact that we are rapidly approaching a point of no return. Meanwhile, the public is becoming more aware of the climate issue and more willing to take active steps to help avert or mitigate the crisis. While current public discourse is mostly centered on how individuals can help slow climate change via small behavioral changes in their everyday lives, such as flying less or not eating meat, the RS approach operates on a much greater scale by targeting major industries. As this Element has explained, RCS does not only target the fossil-fuel industry but also other related or supporting industries, such as finance, advertisement, and retail. Importantly, however, it rejects the notion that the fossil-fuel industry is beyond shaming, by highlighting the great efforts that the industry has invested in building its climate-denial machine.

Such manipulation of public opinion by the fossil-fuel industry, and other carbon-intensive or related industries, is still ongoing and is not a thing of the past. Climate-denial tactics have evolved, and now focus on shifting blame to consumers and on climate delay, by denying the urgency of the situation. Climate-washing practices are also emerging in various consumer and financial contexts. These obstruction tactics are currently a regulatory blind spot (Yadin, forthcoming-b), as regulators tend to focus on direct contributions to climate change in the form of greenhouse gas emissions. Yet, climate obstruction plays a crucial role in the exacerbation of climate change and should be tackled by suitable regulatory means.

Certainly, there is a strong moral argument to be made against the climate-obstruction tactics taken up by the fossil-fuel industry and other industries. This Element, however, is much more practical in nature, focusing not on moral blame (or even legal blame), but instead on how corporate reputational sensitivities, combined with the public's willingness to hold corporations account able for their current (rather than their past) actions, can be used to fight climate change. Thus, the Element presents a practical policy tool, as well as the theory behind it, for encouraging companies to do better on the climate-change front – not only to abide by laws, rules, and regulations but also to adopt voluntary above-compliance norms. While elements of morality are indeed present in the shaming process, RCS is aimed at making corporations more climate-accountable and nudging them to improve, rather than just condemning them for the sake of condemnation, or inflicting punishments or legal remedies such as compensation.

Where can RCS go from here? This Element aimed to lay the groundwork for future research into the conceptual, theoretical, and empirical aspects of RCS. More studies are needed to assess the effectiveness of different policy schemes across jurisdictions, so as to better understand what makes RCS work and what does not. From a legal point of view, it can be valuable to construct legal theories and doctrines that set boundaries for regulatory action and improve fairness toward shamed entities. From a regulatory perspective, it might be useful to study the ways in which shaming affects regulatory institutions, their public image, and their relationships with regulatees, especially compared to other regulatory tools.

While the Element focused on national and subnational regulators, RCS could also be directly deployed by international, supranational, and intergovernmental regulatory bodies such as the EU, the UN, and the OECD. This approach largely falls outside the scope of this project, yet it does warrant further research and could prove useful, especially because carbon majors and other multinational firms that contribute to climate change could be shamed by the larger international community. The architecture of social media also contributes to the potential of climate shaming by international, supranational, and intergovernmental bodies by providing a borderless platform for sharing information and shaming. At a private governance level, private organizations could also adopt climate-shaming policies to advance responsible climate norms within professional associations and industry groups.

At national and subnational levels, it should also be remembered that RS is meant to supplement, and not replace, other regulatory and policy tools.

The challenge facing regulators worldwide is indeed to find the right combination of policy, regulatory, and legal tools that will best balance effectiveness and fairness. Hopefully, shaming methods could help meet this challenge by harnessing corporate reputation concerns, regulatory communications, and public activism toward the goal of slowing climate change and securing our future on this planet.

References

Aaltola, E. (2021). Defensive Over Climate Change? Climate Shame as a Method of Moral Cultivation. *Journal of Agricultural and Environmental Ethics*, 34(1), 5–27. https://link.springer.com/article/10.1007/s10806-021-09844-3#citeas.

Afsah, S., Laplante, B., & Wheeler, D. (1996). Controlling Industrial Pollution: A New Paradigm. *World Bank Policy Research Working Paper No. 1672*. http://www-wds.worldbank.org/external/default/WDSContentServer/WDSP/IB/1996/10/01/000009265_3970311114908/Rendered/PDF/multi_page.pdf.

Ainslie, E. K. (2006). Indicting Corporations Revisited: Lessons of the Arthur Andersen Prosecution. *American Criminal Law Review*, 43(1), 107–142.

Almiron, N., & Xifra, J. (2020). *Climate Change Denial and Public Relations*. London: Routledge.

Ansolabehere, S., & Konisky, D. M. (2014). *Cheap and Clean: How Americans Think about Energy in the Age of Global Warming*. Cambridge, MA: MIT Press.

Ater, I., & Avishay-Rizi, O. (2022). Price Saliency and Fairness: Evidence from Regulatory Shaming. *CEPR Discussion Paper No. DP17156*, https://ssrn.com/abstract=4121331.

Archer, D., & Rahmstorf, S. (2009). *The Climate Crisis: An Introductory Guide to Climate Change*. Cambridge: Cambridge University Press.

Arnold, M. (2022). *ECB Accuses Eurozone Banks of "White Noise" on Climate Risks*, www.ft.com/content/aaa06d90-0356-44b4-b637-0e47c9003ba4.

Ashkenazi, S. (2021). *Yair Avidan: We Will Require Bank Disclosures on Climate Risks and Exposure to Polluting Companies*, www.globes.co.il/news/article.aspx?did=1001395102.

Ayres, I., & Braithwaite, J. (1992). *Responsive Regulation: Transcending the Deregulation Debate*. Oxford: Oxford University Press.

Baldwin, R., Cave, M., & Lodge M. (2012). *Understanding Regulation: Theory, Strategy, and Practice*. Oxford: Oxford University Press.

Bauckloh, T., Klein, C., Pioch, T., & Schiemann, F. (2023). Under Pressure? The Link between Mandatory Climate Reporting and Firms' Carbon Performance. *Organization & Environment*, 36(1), 126–149. https://doi.org/10.1177/10860266221083340.

Bavorova, M., Fietz, A. V., & Hirschauer, N. (2017). Does Disclosure of Food Inspections Affect Business Compliance? The Case of Berlin, Germany. *British Food Journal*, 119(1), 143–163.

Bell, J., Poushter, J., Fagan, M., & Huang, C. (2021). *In Response to Climate Change, Citizens in Advanced Economies Are Willing to Alter How They Live and Work*, www.pewresearch.org/global/2021/09/14/in-response-to-climate-change-citizens-in-advanced-economies-are-willing-to-alter-how-they-live-and-work.

Ben-Shahar, O., & Schneider, C. E. (2014). *More Than You Wanted to Know: The Failure of Mandated Disclosure*. Princeton: Princeton University Press.

Bennear, L. S., & Olmstead, S. M. (2008). The Impacts of the "Right to Know": Information Disclosure and the Violation of Drinking Water Standards. *Journal of Environmental Economics and Management*, 56(2), 117–130.

Benoit, M. (2022). *International Law Obligations on Climate Change Mitigation*. Oxford: Oxford University Press.

Bergquist, P., Konisky, D. M., & Kotcher, J. (2020). Energy Policy and Public Opinion: Patterns, Trends and Future Directions. *Progress in Energy*, 2(3), 032003. https://iopscience.iop.org/article/10.1088/2516-1083/ab9592.

Berliner, D., & Prakash, A. (2013). Signaling Environmental Stewardship in the Shadow of Weak Governance: The Global Diffusion of ISO 14001. *Law & Society Review*, 47(2), 345–373.

Bevan, G., & Hood, C. (2006). What's Measured is What Matters: Targets and Gaming in the English Public Health Care System. *Public Administration*, 84, 517–538.

Bevan, G., & Wilson, D. (2013). Does "Naming and Shaming" Work for Schools and Hospitals? Lessons from Natural Experiments Following Devolution in England and Wales. *Public Money & Management*, 33(4), 245–252.

Bonetti, P., Leuz, C., & Michelon, G. (2023). Internalizing Externalities: Disclosure Regulation for Hydraulic Fracturing, Drilling Activity and Water Quality. European Corporate Governance Institute – Law Working Paper No. 676/2023. https://ssrn.com/abstract=4171246.

Braithwaite, J. (1989). *Crime, Shame, and Reintegration*. Cambridge: Cambridge University Press.

Bratspies, R. M. (2009). Regulatory Trust. *Arizona Law Review*, 51(3), 575–632.

Brady, J., Evans, M. F., & Wehrly, E. W. (2019). Reputational Penalties for Environmental Violations: A Pure and Scientific Replication Study. *International Review of Law and Economics*, 57(C), 60–72.

Branson, D. M. (2002). Corporate Social Responsibility Redux. *Tulane Law Review*, 76(5–6), 1207–1226.

Brooks, J. R., & Ebi, K. L. (2021). Climate Change Warning Labels on Gas Pumps: The Role of Public Opinion Formation in Climate Change Mitigation Policies. *Global Challenges*, 5(10), 2000086. https://onlinelibrary.wiley.com/doi/full/10.1002/gch2.202000086.

Burck, J., Uhlich, T., Bals, C. et al. (2021). *Climate Change Performance Index (CCPI) 2022: Results*. Germanwatch, NewClimate Institute & Climate Action Network. Bonn.

Carlarne, C. P. (2019). U.S. Climate Change Law: A Decade of Flux and an Uncertain Future. *American University Law Review*, 69(2), 387–477.

Carlarne, C. P. (2021). The Essential Role of Climate Litigation and the Courts in Averting Climate Crisis. In B. Mayer & A. Zahar, eds., *Debating Climate Law*. Cambridge: Cambridge University Press, pp. 111–127.

Carlarne, C. P., Gray, K. R., & Tarasofsky, R. (2016). International Climate Change Law: Mapping the Field. In K. R. Gray, R. Tarasofsky, & C. P. Carlarne, eds., *The Oxford Handbook of International Climate Change Law*. Oxford University Press (Oxford Handbooks Online), pp. 3–25.

CAT. (2021). *Glasgow's 2030 Credibility Gap: Net Zero's Lip Service to Climate Action*, https://climateactiontracker.org/publications/glasgows-2030-credibility-gap-net-zeros-lip-service-to-climate-action.

Choudhury, B. (2021). Climate Change as Systemic Risk. *Berkeley Business Law Journal*, 18(2), 52–93.

Christensen, H. B., Floyd, E., Liu, L. Y., & Maffett, M. (2017). The Real Effects of Mandated Information on Social Responsibility in Financial Reports: Evidence from Mine-Safety Records. *Journal of Accounting and Economics*, 64(2), 284–304.

Coen, D., Kreienkamp, J., & Pegram, T. (2020). *Global Climate Governance* (Elements in Public and Nonprofit Administration). Cambridge: Cambridge University Press.

Cohen, M. A., & Viscusi, W. K. (2012). The Role of Information Disclosure in Climate Mitigation Policy. *Climate Change Economics*, 3(4), 1–21.

COP26. (2021). *COP26: The Glasgow Climate Pact*, https://ukcop26.org/the-conference/cop26-outcomes.

Cortez, N. (2011). Adverse Publicity by Administrative Agencies in the Internet Era. *Brigham Young University Law Review*, 2011(5), 1371–1454.

Cortez, N. (2018). Regulation by Database. *University of Colorado Law Review*, 89(1), 1–92.

Dadush, S. (2018). Identity Harm. *University of Colorado Law Review*, 89(3), 863–936.

Daly, P. (2021). *Understanding Administrative Law in the Common Law World*. Oxford: Oxford University Press.

Darko-Mensah, A., & Okereke, C. (2013). Can Environmental Performance Rating Programs Succeed in Africa? An Evaluation of Ghana's AKOBEN Project. *Management of Environmental Quality*, 25(5), 599–618.

Delmas, M., Montes-Sancho, M. J., & Shimshack, J. P. (2010). Information Disclosure Policies: Evidence from the Electricity Industry. *Economic Inquiry*, 48(2), 483–498.

Dessler, A. E. (2021). *Introduction to Modern Climate Change*, 3rd ed. Cambridge: Cambridge University Press.

Dessler, A. E., & Parson, E. A. (2019). *The Science and Politics of Global Climate Change: A Guide to the Debate*. Cambridge: Cambridge University Press.

Devine, D., Gaskell, J., Jennings, W., & Stoker, G. (2020). Trust and the Coronavirus Pandemic: What Are the Consequences of and for Trust? An Early Review of the Literature. *Political Studies Review*, 19(2), 274–285.

DiMaggio, P. J., & Powell, W. W. (1983). The Iron Cage Revisited: Institutional Isomorphism and Collective Rationality in Organizational Fields. *American Sociological Review*, 48(2), 147–160.

Doonan, J., Lanoie, P., & Laplante, B. (2002). Environmental Performance of Canadian Pulp and Paper Plants: Why Some Do Well and Others Do Not? *Cahiers de Recherche* 02–01, HEC Montréal, Institut d'économie appliquée. (2002s–24, Working Papers, CIRANO). https://cirano.qc.ca/en/summaries/2002s-24.

Downar, B., Ernstberger, J., Reichelstein, S., Schwenen, S., & Zaklan, A. (2021). The Impact of Carbon Disclosure Mandates on Emissions and Financial Operating Performance. *Review of Accounting Studies*, 26, 1137–1175.

Egan, P., Konisky, D., & Mullin, M. (2022). Ascendant Public Opinion: The Rising Influence of Climate Change on Americans' Attitudes about the Environment. *Public Opinion Quarterly*, 86(1), 134–148.

Eichhorn, J., Molthof, L., & Nicke, S. (2020). *From Climate Change Awareness to Climate Crisis Action: Public Perceptions in Europe and the United States*. New York: Open Society Foundations.

Eskander, S. M., & Fankhauser, S. (2020). Reduction in Greenhouse Gas Emissions from National Climate Legislation. *Nature Climate Change*, 10, 750–756.

Etzioni, A. (2003). *The Monochrome Society*. Princeton: Princeton University Press.

European Commission. (n.d.-a). *EU Emissions Trading System (EU ETS)*, https://ec.europa.eu/clima/eu-action/eu-emissions-trading-system-eu-ets_en.

European Commission. (n.d.-b). *Tyres: Energy Labelling Requirements Apply to this Product*, https://ec.europa.eu/info/energy-climate-change-environment/standards-tools-and-labels/products-labelling-rules-and-requirements/energy-label-and-ecodesign/energy-efficient-products/tyres_en.

European Environment Agency (2022). *Climate Change Mitigation Policy and Measures (Greenhouse Gas Emission)*, www.eea.europa.eu/data-and-maps/data/climate-change-mitigation-policies-and-measures-1.

Fancourt, D., Steptoe, A., & Wright, L. (2020). The Cummings Effect: Politics, Trust, and Behaviours During the COVID-19 Pandemic. *The Lancet*, 396 (10249), 464–465.

Fankhauser, S., Hepburn, C., & Park, J. (2010). Combining Multiple Climate Policy Instruments: How Not to Do It. *Climate Change Economics*, 1(3), 209–225.

Fischedick M., Roy, J., Abdel-Aziz, A., Acquaye, A. et al., eds., *Climate Change 2014: Mitigation of Climate Change*. Cambridge: Cambridge University Press.

Fleurke, F., & Verschuuren, J. (2016). Enforcing the European Emissions Trading System within the EU Member States: A Procrustean Bed? In R. White, T. Spapens, & W. Huisman, eds., *Environmental Crime in Transnational Context: Global Issues in Green Enforcement and Criminology*. London: Routledge, pp. 208–230.

Fransen, T. (2021). *Making Sense of Countries' Paris Agreement Climate Pledges*, www.wri.org/insights/understanding-ndcs-paris-agreement-climate-pledges.

Fredericks, S. E. (2021). *Environmental Guilt and Shame: Signals of Individual and Collective Responsibility and the Need for Ritual Responses*. Oxford: Oxford University Press.

Freeman, J. (2020). The Environmental Protection Agency's Role in US Climate Policy: A Fifty Year Appraisal. *Duke Environmental Law and Policy Forum*, 31, 1–79.

Freeman, J. (2021). EPA and Climate Change. In A. James Barnes, J. D. Graham, & D. M. Konisky, eds., *Fifty Years at the US Environmental Protection Agency*. London: Rowman & Littlefield, pp. 121–165.

Freese, B. (2020). *Industrial-Strength Denial: Eight Stories of Corporations Defending the Indefensible, from the Slave Trade to Climate Change*. Oakland: University of California Press.

Frumhoff, P. C., Heede, R., & Oreskes, N. (2015). The Climate Responsibilities of Industrial Carbon Producers. *Climatic Change*, 132, 157–171.

Fung, A., Graham, M., & Weil, D. (2007). *Full Disclosure: The Perils and Promise of Transparency*. Cambridge: Cambridge University Press.

Fung, A., & O'Rourke, D. (2000). FORUM: Reinventing Environmental Regulation from the Grassroots Up: Explaining and Expanding the Success of the Toxics Release Inventory. *Environmental Management*, 25(2), 115–127.

Future Earth, The Earth League & WCRP. (2022). *10 New Insights in Climate Science 2022*. Stockholm. https://10insightsclimate.science.

Garvey, S. P. (1998). Can Shaming Punishments Educate? *University of Chicago Law Review*, 65(3), 733–794.

Gee, M. L., & Copeland, D. (2022). Shaming: A Concept Analysis. *Advances in Nursing Science*, 10(1097), 1–13.

Geiger, N., Gore, A., Squire, C. V. & Attari, S. Z. (2021). Investigating Similarities and Differences in Individual Reactions to the COVID-19 Pandemic and the Climate Crisis. *Climatic Change*, 167(1), 1–20. https://link.springer.com/article/10.1007/s10584-021-03143-8.

Genovese, F. (2020). *Weak States at Global Climate Negotiations* (Elements in International Relations). Cambridge: Cambridge University Press.

Godfrey, P. C., Merrill, C. B., & Hansen, J. M. (2009). The Relationship between Corporate Social Responsibility and Shareholder Value: An Empirical Test of the Risk Management Hypothesis. *Strategic Management Journal*, 30, 425–445.

Gunningham, N., Grabosky, P., & Sinclair, D. (1998). *Smart Regulation: Designing Environmental Policy*. Oxford: Oxford University Press.

Gunningham, N., Kagan, R. A., & Thornton, D. (2003). *Shades of Green: Business, Regulation, and Environment*. Stanford: Stanford Law and Politics.

Gunningham, N., Kagan, R. A., & Thornton, D. (2004). Social License and Environmental Protection: Why Businesses Go beyond Compliance. *Law & Social Inquiry*, 29(2), 307–341.

Gunningham, N., & Sinclair, D. (2017). Smart Regulation. In P. Drahos, ed., *Regulatory Theory: Foundations and Applications*. Australia: ANU Press, 133–148.

Gupta, S., & Goldar, B. (2005). Do Stock Markets Penalize Environment-Unfriendly Behaviour? Evidence from India. *Ecological Economics*, 52(1), 81–95.

Gupta, S., Tirpak, D. A., Burger, N. et al. (2007). Policies, Instruments and Co-operative Arrangements. In B. Metz, et al. eds., *Climate Change 2007: Mitigation*. Cambridge: Cambridge University Press, 745–807.

Hagan, S., & Brush, S. (2022). *Texas Republicans Squeeze Wall Street Firms over Climate Policies*, www.latimes.com/business/story/2022-04-26/texas-republicans-push-back-on-wall-streets-embrace-of-climate-policies?.

Haines, F., & Parker, C. (2017). Moving towards Ecological Regulation: The Role of Criminalisation. In C. Holley & C. Shearing, eds., *Criminology and the Anthropocene*. Abingdon, UK: Routledge – Taylor & Francis, 81–108.

Heede, R. (2014). Tracing Anthropogenic Carbon Dioxide and Methane Emissions to Fossil Fuel and Cement Producers, 1854–2010. *Climatic Change*, 122, 229–241.

Heede, R. (2020). Carbon Majors 2018. Data set released December 2020, Climate Accountability Institute, Colorado USA. https://climateaccountabil ity.org/carbonmajors.html (last accessed March 24, 2023).

Hertwich, E. G., & Wood, R. (2018). The Growing Importance of Scope-3 Greenhouse Gas Emissions from Industry. *Environmental Research Letters*, 13, 104013. https://iopscience.iop.org/article/10.1088/1748-9326/aae19a.

Horton, H. (2022). *Greenwashing UK Fashion Firms to be Named and Shamed by Watchdog*, www.theguardian.com/fashion/2022/mar/11/greenwashing-uk-fashion-firms-to-be-named-and-shamed-by-watchdog.

Hsueh, L. (2020). Calling all Volunteers: Industry Self-Regulation on the Environment. In D. Konisky, ed., *Handbook of U.S. Environmental Policy*. Gloucestershire, UK: Edward Elgar, 243–256.

Hsueh, L., & Prakash, A. (2012). Incentivizing Self-Regulation: Federal vs. State-Level Voluntary Programs in US Climate Change Policies. *Regulation & Governance*, 6(4), 445–473.

Huang, A. H., Shen, M., Tang, C., & Wang, J. (2022). The Effects of Regulatory Enforcement Disclosure: Evidence from OSHA's Press Release about Safety Violations. *Working Paper*.

Huang, J. (2021). Exploring Climate Framework Laws and the Future of Climate Action. *Pace Environmental Law Review*, 38(2), 285–326.

Huggins, A. (2021). The Paris Agreement's Article 15 Mechanism: An Incomplete Compliance Strategy. In B. Mayer & A. Zahar, eds., *Debating Climate Law*. Cambridge: Cambridge University Press, pp. 99–110.

Hughes, L., Konisky, D. M., & Potter, S. (2020). Extreme Weather and Climate Opinion: Evidence from Australia. *Climatic Change*, 163, 723–743.

IEA. (2021). *World Energy Outlook 2021*, www.iea.org/reports/world-energy-outlook-2021.

IPCC. (2021). *Climate Change 2021: The Physical Science Basis*. https://www .ipcc.ch/report/ar6/wg1.

IPCC. (2022a). *Climate Change 2022: Impacts, Adaptation and Vulnerability*. https://www.ipcc.ch/report/ar6/wg2.

IPCC. (2022b). *Climate Change 2022: Mitigation of Climate Change*. https:// www.ipcc.ch/report/ar6/wg3.

Israeli Ministry of Energy. (2020). *Ministry of Energy Announces: For the First Time in Israel, Energy Rating for New Homes*, www.gov.il/en/departments/ news/press_230620.

Israeli Ministry of Environmental Protection. (n.d.). *Reporting on Greenhouse Gas Emissions*, www.gov.il/en/Departments/Guides/reporting_on_greenhouse_gas_emissions.

Jacquet, J. (2015). *Is Shame Necessary? New Uses for An Old Tool*. New York: Knopf Doubleday.

Jacquet, J., & Jamieson, D. (2016). Soft but Significant Power in the Paris Agreement. *Nature Climate Change*, 6, 643–646.

Jin, G. Z., & Leslie, P. (2003). The Effect of Information on Product Quality: Evidence from Restaurant Hygiene Grade Cards. *Quarterly Journal of Economics*, 118(2), 409–451.

Johnson, M. S. (2020). Regulation by Shaming: Deterrence Effects of Publicizing Violations of Workplace Safety and Health Laws. *American Economic Review*, 110(6), 1866–1904.

Jouvenot, V., & Krueger, P. (2019). Mandatory Corporate Carbon Disclosure: Evidence from a Natural Experiment. *SSRN*, August 13. http://dx.doi.org/10.2139/ssrn.3434490.

Kahan, D. M. (1996). What Do Alternative Sanctions Mean? *University of Chicago Law Review*, 63(2), 591–653.

Kahan, D. M. (2006). What's Really Wrong with Shaming Sanctions. *Texas Law Review*, 84(7), 2075–2096.

Karpoff, J. M. (2012). Does Reputation Work to Discipline Corporate Misconduct? In M. Barnett & T. G. Pollock, eds., *The Oxford Handbook of Corporate Reputation*. Oxford: Oxford University Press, pp. 361–382.

Karpoff, J. M., Lott, Jr., J. R., & Wehrly, E. W. (2005). The Reputational Penalties for Environmental Violations: Empirical Evidence. *The Journal of Law & Economics*, 48(2), 653–675.

Klonick, K. (2016). Re-Shaming the Debate: Social Norms, Shame, and Regulation in an Internet Age. *Maryland Law Review*, 75(4), 1029–1065.

Konisky, D. M., Hughes, L., & Kaylor, C. H. (2016). Extreme Weather Events and Climate Change Concern. *Climatic Change*, 134, 533–547.

Konisky, D. M., & Woods, N. D. (2016). Environmental Policy, Federalism, and the Obama Presidency. *Publius: The Journal of Federalism*, 46(3), 366–391.

Konisky, D. M., & Woods, N. D. (2018). Environmental Federalism and the Trump Presidency: A Preliminary Assessment. *Publius: The Journal of Federalism*, 48(3), 345–371.

Koop, C., & Lodge, M. (2017). What Is Regulation? An Interdisciplinary Concept Analysis. *Regulation & Governance*, 11(1), 95–108.

Kramer, R. C. (2020). *Carbon Criminals, Climate Crimes*. New Brunswick: Rutgers University Press.

Laffont, J. J., & Martimort, D. (1999). Separation of Regulators against Collusive Behavior. *The Rand Journal of Economics*, 30(2), 232–262.

Lamb, S. (2003). The Psychology of Condemnation: Underlying Emotions and Their Symbolic Expression in Condemning and Shaming. *Brooklyn Law Review*, 68(4), 929–958.

Le Quéré, C., Peters, G. P., Friedlingstein, P. et al. (2021). Fossil CO_2 Emissions in the Post-COVID-19 Era. *Nature Climate Change*, 11, 197–199.

Leiserowitz, A. et al. (2019). *Climate Change in the American Mind: November 2019*. New Haven: Yale Program on Climate Change Communication.

Leiserowitz, A. et al. (2021a). *Climate Change in the American Mind: September 2021*. New Haven: Yale Program on Climate Change Communication.

Leiserowitz, A., et al. (2021b). *Consumer Activism on Global Warming: September 2021*. New Haven: Yale Program on Climate Change Communication.

Leiserowitz, A., Carman, J., Rosenthal, S. et al. (2021c). *Climate Change in the Irish Mind*. Yale Program on Climate Change Communication and the Irish Environmental Protection Agency.

Leiserowitz, A., Maibach, E., Rosenthal, S. et al. (2022). *Politics & Global Warming*. Yale Program on Climate Change Communication.

Li, M., Trencher, G., & Asuka, J. (2022). The Clean Energy Claims of BP, Chevron, ExxonMobil and Shell: A Mismatch between Discourse, Actions and Investments. *PloS ONE*, 17(2), e0263596. https://doi.org/10.1371/jour nal.pone.0263596.

Loewenstein, G., Sunstein, C. R., & Golman, R. (2014). Disclosure: Psychology Changes Everything. *Annual Review of Economics*, 6, 391–419.

Luna, J. C. (2019). *Naming and Shaming in the European Union Emission Trading Scheme: A Legal Review*. Thesis. Maastricht, the Netherlands: Maastricht University.

Lyster, R. (2016). *Climate Justice and Disaster Law*. Cambridge: Cambridge University Press.

Lyster, R. (2021). Climate Change Law (2019). *Yearbook of International Disaster Law Online*, 2(1), 450–462.

Maslin, M. (2021). *Climate Change: A Very Short Introduction*, 4th ed. Oxford: Oxford University Press.

Massaro, T. M. (1991). Shame, Culture, and American Criminal Law. *Michigan Law Review*, 89(7), 1880–1944.

Mayer, B. (2021). Reflection 1: Adaptation. In B. Mayer & A. Zahar, eds., *Debating Climate Law*. Cambridge: Cambridge University Press, pp. 310–328.

McDonald, J., & McCormack, P. C. (2021). Rethinking the Role of Law in Adapting to Climate Change. *WIREs Climate Change*, 12(5), e726. https://wires.onlinelibrary.wiley.com/doi/10.1002/wcc.726.

McDonald, M. (2021). *Ecological Security: Climate Change and the Construction of Security*. Cambridge: Cambridge University Press.

McGreal, C. (2021). *Revealed: 60% of Americans Say Oil Firms Are to Blame for the Climate Crisis*, www.theguardian.com/environment/2021/oct/26/climate-change-poll-oil-gas-companies-environment.

McGuire, W., Holtmaat, E. A., & Prakash, A. (2022). Penalties for Industrial Accidents: The Impact of the Deepwater Horizon Accident on BP's Reputation and Stock Market Returns. *PloS one*, 17(6), e0268743. https://doi.org/10.1371/journal.pone.0268743.

Meckling, J., & Allan, B. B. (2020). The Evolution of Ideas in Global Climate Policy. *Nature Climate Change*, 10, 434–438.

Meijer, A. (2013). Local Meanings of Targeted Transparency: Understanding the Fuzzy Effects of Disclosure Systems. *Administrative Theory & Praxis*, 35(3), 398–423.

Meijer, A., & Homburg, V. (2009). Disclosure and Compliance: The "Pillory" as an Innovative Regulatory Instrument. *Information Polity*, 14(4), 279–294.

Michaels, D. (2008). *Doubt Is Their Product: How Industry's Assault on Science Threatens Your Health*. Oxford: Oxford University Press.

Michaels, D. (2020). *The Triumph of Doubt: Dark Money and the Science of Deception*. Oxford: Oxford University Press.

Mkono, M., & Hughes, K. (2020). Eco-Guilt and Eco-Shame in Tourism Consumption Contexts: Understanding the Triggers and Responses. *Journal of Sustainable Tourism*, 28(8), 1223–1244.

Moffa, A. (2020). Uniform Climate Control. *University of Richmond Law Review*, 54(4), 993–1044.

Moss, J., & Fraser, P. (2019) *Australia's Carbon Majors*. Sydney: UNSW Practical Justice Initiative.

Nussbaum, M. C. (2004). *Hiding from Humanity: Disgust, Shame, and the Law*. Princeton: Princeton University Press.

OSHA. (2022). *Commonly Used Statistics*, www.osha.gov/data/commonstats.

Potoski, M., & Prakash, A. (2009). *Voluntary Programs: A Club Theory Perspective*. Cambridge, MA: MIT Press.

Poushter, J., & Huang, C. (2019). *Climate Change Still Seen as the Top Global Threat, but Cyberattacks a Rising Concern*, www.pewresearch.org/global/2019/02/10/climate-change-still-seen-as-the-top-global-threat-but-cyberattacks-a-rising-concern/#table.

Prakash, A. (2000). *Greening the Firm: The Politics of Corporate Environmentalism*. Cambridge: Cambridge University Press.

Prakash, A., & Potoski, M. (2006). *The Voluntary Environmentalists: Green Clubs, ISO 14001, and Voluntary Environmental Regulations*. Cambridge: Cambridge University Press.

Reisch, L., Sunstein, C. R., & Kaiser, M. (2021). What Do People Want to Know? Information Avoidance and Food Policy Implications. *Food Policy*, 102(102076). https://doi.org/10.1016/j.foodpol.2021.102076.

Scotford, E., Minas, S., & Macintosh, A. (2017). Climate Change and National Laws across Commonwealth Countries. *Commonwealth Law Bulletin*, 43(3–4), 318–361.

Shapira, R. (2022). The Challenge of Holding Big Business Accountable. *Cardozo Law Review*, 44(1), 203–270.

Shapiro, M. A. (2020). The Indignities of Civil Litigation. *Boston Law Review*, 100(2), 501–579.

Shimshack, J. P. (2020). Information Provision. In D. Konisky, ed., *Handbook of US Environmental Policy*. Gloucestershire, UK: Edward Elgar, pp. 231–242.

Simon, P. A. et al. (2005). Impact of Restaurant Hygiene Grade Cards on Foodborne-Disease Hospitalizations in Los Angeles County. *Journal of Environmental Health*, 67(7), 32–60.

Skeel, D. A. (2001). Shaming in Corporate Law. *University of Pennsylvania Law Review*, 149(6), 1811–1866.

Solove, D. J. (2007). The *Future of Reputation: Gossip, Rumor, and Privacy on the Internet*. New Haven: Yale University Press.

Sørensen, E., & Torfing, J. (2022). Co-Creating Ambitious Climate Change Mitigation Goals: The Copenhagen Experience. *Regulation & Governance*, 16(2), 572–587.

Spektor, M., Mignozzetti, U., & Fasolin, G. (2022). Nationalist Backlash against Foreign Climate Shaming. *Global Environmental Politics*, 22(1), 139–158.

Stephan, M. (2002). Environmental Information Disclosure Programs: They Work, But Why? *Social Science Quarterly*, 83(1), 190–205.

Stiglitz, E. H. (2018). Delegating for Trust. *University of Pennsylvania Law Review*, 166(3), 633–698.

Sunstein, C. R. (2020). *Too Much Information: Understanding What You Don't Want to Know*. Cambridge, MA: MIT Press.

Supran, G., & Oreskes, N. (2021). Rhetoric and Frame Analysis of ExxonMobil's Climate Change Communications. *One Earth*, 4(5), 696–719.

Swedish Energy Agency. (2021). STEMFS 2021:5 – Regulations Amending the Swedish Energy Agency's Regulations on the Obligation to Provide

Consumers with Environmental Information on Fuels. https://energimyndigh eten.a-w2m.se/Home.mvc?ResourceId=183767.

Taebi, B., & Safari, A. (2017). On Effectiveness and Legitimacy of "Shaming" as a Strategy for Combatting Climate Change. *Science and Engineering Ethics*, 23, 1289–1306.

Tangney, J. P., Burggraf, S. A., & Wagner, P. E. (1995). Shame-Proneness, Guilt-Proneness, and Psychological Symptoms. In J. P. Tangney & K. W. Fischer, eds., *Self-Conscious Emotions: The Psychology of Shame, Guilt, Embarrassment, and Pride*. New York: Guilford Press, pp. 343–367.

Taufique, K. M. R., Nielsen, K. S., Dietz, T. et al. (2022). Revisiting the Promise of Carbon Labelling. *Nature Climate Change*, 12, 132–140,

Teichman, D., & Zamir, E. (2022). Exponential Growth Bias and the Law: Why Do We Save Too Little, Borrow Too Much, and Fail to React on Time to Deadly Pandemics and Climate Change? *Vanderbilt Law Review*, 75(5), 1345–1400.

Tingley, D., & Tomz, M. (2022). The Effects of Naming and Shaming on Public Support for Compliance with International Agreements: An Experimental Analysis of the Paris Agreement. *International Organization*, 76, 445–468.

Tomar, S. (2022). Greenhouse Gas Disclosure and Emissions Benchmarking. *SMU Cox School of Business Research Paper No. 19–17*. http://dx.doi.org/ 10.2139/ssrn.3448904.

UNEP. (2021). *Emissions Gap Report 2021*. www.unep.org/resources/emis sions-gap-report-2021.

UNEP. (2022). *Adaptation Gap Report 2022*. www.unep.org/resources/adapta tion-gap-report-2022.

UNFCCC. (2021). *Secretary-General's Statement on the IPCC Working Group 1 Report on the Physical Science Basis of the Sixth Assessment*, https:// unfccc.int/news/secretary-general-s-statement-on-the-ipcc-working-group- 1-report-on-the-physical-science-basis-of.

Van der Zee, B., & Horton, H. (2022). *Cop27 Day One: UN Chief Warns World Is "On Highway to Climate Hell" – As It Happened*, https://www.theguar dian.com/environment/live/2022/nov/07/cop27-egypt-climate-summit- boris-johnson-net-zero-live.

Van Erp, J. (2007). Effects of Disclosure on Business Compliance: A Framework for the Analysis of Disclosure Regimes. *European Food and Feed Law Review*, 2, 255–263.

Van Erp, J. (2010). Regulatory Disclosure of Offending Companies in the Dutch Financial Market: Consumer Protection or Enforcement Publicity? *Law & Policy*, 32(4), 407–433.

Van Erp, J. (2011). Naming without Shaming: The Publication of Sanctions in the Dutch Financial Market. *Regulation & Governance*, 5(3), 287–308.

Van Erp, J. (2021). Shaming and Compliance. In B. Van Rooij & D. Sokol, eds., *The Cambridge Handbook of Compliance*. Cambridge: Cambridge University Press, pp. 438–450.

Washington, H., & Cook, J. (2011). *Climate Change Denial: Heads in the Sand*. London: Routledge.

Wennersten, J. R., & Robbins, D. (2017). *Rising Tides: Climate Refugees in the Twenty-First Century*. Bloomington: Indiana University Press.

Whitman, J. Q. (1998). What Is Wrong with Inflicting Shame Sanctions? *Yale Law Journal*, 107(5), 1055–1092.

Williams, K. D. (2007). Ostracism: The Kiss of Social Death. *Social and Personality Psychology Compass*, 1(1), 236–247.

Wilson, D. (2004). Which Ranking? The Impact of a "Value-Added" Measure of Secondary School Performance. *Public Money and Management*, 24(1), 37–45.

WMO. (2021a). *State of Climate in 2021: Extreme Events and Major Impacts*, https://public.wmo.int/en/media/press-release/state-of-climate-2021-extreme-events-and-major-impacts.

WMO. (2021b). *Weather-Related Disasters Increase Over Past 50 Years, Causing More Damage but Fewer Deaths*, https://public.wmo.int/en/media/press-release/weather-related-disasters-increase-over-past-50-years-causing-more-damage-fewer.

WMO. (2022). *WMO Update: 50:50 Chance of Global Temperature Temporarily Reaching 1.5°C Threshold in Next Five Years*, https://public.wmo.int/en/media/press-release/wmo-update-5050-chance-of-global-temperature-temporarily-reaching-15%C2%B0c-threshold.

WMO. (2023). *Past Eight Years Confirmed to Be the Eight Warmest on Record*, https://public.wmo.int/en/media/press-release/past-eight-years-confirmed-be-eight-warmest-record.

World Bank. (2020). *World Bank Reference Guide to Climate Change Framework Legislation*. Washington, DC: World Bank.

Yadin, S. (2019a). Regulatory Shaming. *Environmental Law*, 49(2), 407–451.

Yadin, S. (2019b). Saving Lives through Shaming. *Harvard Business Law Review Online*, 9, 57–68.

Yadin, S. (2019c). Shaming Big Pharma. *Yale Journal on Regulation Bulletin*, 36, 131–147.

Yadin, S. (2020). E-Regulation. *Cardozo Arts & Entertainment Law Journal*, 38 (1), 101–152.

Yadin, S. (2021a). Israel's Law and Regulation after the Gas Discoveries. In E. Tevet, V. Shiffer, & I. Galnoor, eds., *Regulation in Israel*. Cham: Palgrave Macmillan, pp. 217–238.

Yadin, S. (2021b). Manipulating Disclosure: Creative Compliance in the Israeli Food Industry. *Saint Louis University Law Journal*, 66(1), 149–166.

Yadin, S. (forthcoming-a). Government Regulation by Eco-Shaming Corporations: Balancing Effectiveness and Fairness. In M. Pinto & G. Seidman, eds., *The Legal Aspects of Shaming: An Ancient Sanction in the Modern World*. Gloucestershire, UK: Edward Elgar. https://papers.ssrn .com/sol3/papers.cfm?abstract_id=4460874.

Yadin, S. (forthcoming-b). Regulatory Shaming and the Problem of Corporate Climate Obstruction. *Harvard Journal on Legislation*, 60.

Yadin, S. (forthcoming-c). The Crowdsourcing of Regulatory Monitoring and Enforcement. *Law and Ethics of Human Rights*, 17(1). https://papers.ssrn .com/sol3/papers.cfm?abstract_id=4441862.

Yang, L., Muller, N. Z., & Liang P. J. (2021). The Real Effects of Mandatory CSR Disclosure on Emissions: Evidence from the Greenhouse Gas Reporting Program, *NBER Working Paper No. 28984*. www.nber.org/papers/w28984.

Acknowledgments

I thank the Max Stern Yezreel Valley College and the Minerva Center for the Rule of Law under Extreme conditions, Faculty of Law and Department of Geography and Environmental Studies, University of Haifa for supporting this Element. I also thank the series editors, Professors Aseem Prakash and David Konisky for their valuable input and their thoughtful assistance and advice throughout this project and the anonymous reviewers for their helpful suggestions. Parts of this project were presented in various forums, including a conference for judges on environmental protection, human rights, and the climate crisis; the Public Policy and Behavioral Change Lecture Series, hosted by the Hebrew University of Jerusalem School of Public Policy and Governance; the Corporate Responsibility and Liability in Relation to Climate Change Conference, hosted by Utrecht University; the Yezreel Valley College School of Public Administration and Public Policy Colloquium; the Israel Political Science Association Annual Conference; the Israeli Law and Society Association Annual Conference; and the Annual Science and Environment Conference, hosted by the Israeli Society of Ecology and Environmental Sciences. I thank all participants for their useful comments. Special thanks go to Tamar Yadin and Daniel Barnett for their technical assistance. Finally, I thank my beloved husband Gilad, for his endless encouragement and support.

About the Author

Dr. Sharon Yadin is an associate professor of law and regulation at the Yezreel Valley College School of Public Administration and Public Policy. She has published some thirty articles in prominent journals in Israel and the United States, including *Harvard Journal on Legislation, Environmental Law, Yale Journal on Regulation Bulletin,* and *Harvard Business Law Review Online,* as well as two previous books. Dr. Yadin's research focuses on soft regulatory approaches of government agencies. Her work on regulatory contracts was adopted into law in *MQG* v. *Prime Minister of Israel,* a precedential Supreme Court of Israel ruling on natural gas regulation in which her first book was cited more than a dozen times in four different opinions. Her studies on regulatory shaming were cited in leading publications in the field, presented in prominent international fora, and have influenced policy in Israel and abroad. Dr. Yadin earned her doctorate degree in law from Tel Aviv University and completed her post-doctorate at the Hebrew University of Jerusalem. She has won several prestigious scholarships, research grants, and academic awards. She serves on public boards and committees and advises regulators and corporations on legislative and regulatory matters.

Cambridge Elements ≡

Organizational Response to Climate Change

Aseem Prakash

University of Washington

Aseem Prakash is Professor of Political Science, the Walker Family Professor for the College of Arts and Sciences, and the Founding Director of the Center for Environmental Politics at University of Washington, Seattle. His recent awards include the American Political Science Association's 2020 Elinor Ostrom Career Achievement Award in recognition of "lifetime contribution to the study of science, technology, and environmental politics," the International Studies Association's 2019 Distinguished International Political Economy Scholar Award that recognizes "outstanding senior scholars whose influence and path-breaking intellectual work will continue to impact the field for years to come," and the European Consortium for Political Research Standing Group on Regulatory Governance's 2018 Regulatory Studies Development Award that recognizes a senior scholar who has made notable "contributions to the field of regulatory governance".

Jennifer Hadden

University of Maryland

Jennifer Hadden is Associate Professor in the Department of Government and Politics at the University of Maryland. She conducts research in international relations, environmental politics, network analysis, non-state actors and social movements. Her research has been published in various journals, including the *British Journal of Political Science, International Studies Quarterly, Global Environmental Politics, Environmental Politics,* and *Mobilization.* Dr. Hadden's award-winning book, *Networks in Contention: The Divisive Politics of Global Climate Change,* was published by Cambridge University Press in 2015. Her research has been supported by a Fulbright Fellowship, as well as grants from the National Science Foundation, the National Socio-Environmental Synthesis Center, and others. She held an International Affairs Fellowship from the Council on Foreign Relations for the 2015–2016 academic year, supporting work on the Paris Climate Conference in the Office of the Special Envoy for Climate Change at the U.S. Department of State.

David Konisky

Indiana University

David Konisky is Professor at the Paul H. O'Neill School of Public and Environmental Affairs, Indiana University, Bloomington. His research focuses on U.S. environmental and energy policy, with particular emphasis on regulation, federalism and state politics, public opinion, and environmental justice. His research has been published in various journals, including the *American Journal of Political Science, Climatic Change, the Journal of Politics, Nature Energy,* and *Public Opinion Quarterly.* He has authored or edited six books on environmental politics and policy, including *Fifty Years at the U.S. Environmental Protection Agency: Progress, Retrenchment and Opportunities* (Rowman & Littlefield, 2020, with Jim Barnes and John D. Graham), *Failed Promises: Evaluating the Federal Government's Response to Environmental Justice* (MIT Press, 2015), and *Cheap and Clean: How Americans Think about Energy in the Age of Global Warming* (MIT Press, 2014, with Steve Ansolabehere). Konisky's research has been funded by the National Science Foundation, the Russell Sage Foundation, and the Alfred P. Sloan Foundation. Konisky is currently co-editor of *Environmental Politics.*

Matthew Potoski

UC Santa Barbara

Matthew Potoski is a Professor at UCSB's Bren School of Environmental Science and Management. He currently teaches courses on corporate environmental management, and his research focuses on management, voluntary environmental programs, and public policy. His research has appeared in business journals such as *Strategic Management Journal, Business Strategy and the Environment, and the Journal of Cleaner Production,* as well as public policy and management journals such as *Public Administration Review* and *the Journal of Policy Analysis and Management.* He co-authored *The Voluntary Environmentalists* (Cambridge, 2006) and *Complex Contracting* (Cambridge, 2014; the winner of the 2014 Best Book Award, American Society for Public Administration, Section on Public Administration Research) and was co-editor of *Voluntary Programs* (MIT, 2009). Professor Potoski is currently co-editor of the *Journal of Policy Analysis and Management* and the *International Public Management Journal.*

About the Series

How are governments, businesses, and non-profits responding to the climate challenge in terms of what they do, how they function, and how they govern themselves? This series seeks to understand why and how they make these choices and with what consequence for the organization and the eco-system within which it functions.

Cambridge Elements ☰

Organizational Response to Climate Change

Elements in the Series

*Explaining Transformative Change in ASEAN and EU Climate Policy: Multilevel
Problems, Policies and Politics*
Charanpal Bal, David Coen, Julia Kreienkamp, Paramitaningrum and Tom Pegram

Fighting Climate Change through Shaming
Sharon Yadin

A full series listing is available at: www.cambridge.org/ORCC

Printed in the United States
by Baker & Taylor Publisher Services